TRICKLE DOWN TYRANNY

ALSO BY MICHAEL SAVAGE

NONFICTION

Trickle Up Poverty
Banned in Britain
Psychological Nudity
The Political Zoo
Liberalism Is a Mental Disorder
The Enemy Within
The Savage Nation
The Death of the White Male

FICTION

Abuse of Power

TRICKLE DOWN TYRANNY

CRUSHING OBAMA'S DREAM OF THE SOCIALIST STATES OF AMERICA

MICHAEL SAVAGE

wm

WILLIAM MORROW

An Imprint of HarperCollins*Publishers*

HarperCollins books may be purchased for educational, business, or sales promotional use. For information please write: Special Markets Department, HarperCollins Publishers, 10 East 53rd Street, New York, NY 10022.

FIRST EDITION

Library of Congress Cataloging-in-Publication Data has been applied for.

ISBN 978-0-06-208397-5

12 13 14 15 16 DIX/BVG 10 9 8 7 6 5 4 3 2 1

This book is dedicated to those who gave their lives to secure the freedom to write and read as we please

CONTENTS

TRICKLE
DOWN
TYRANNY

CHAPTER 1

Advice to the Next President: A Savage Worldview

Economies can be rebuilt, armies can be repopulated, but once a nation's pride is gone it can almost never be restored. The loss of a nation's honor is something not even centuries can repair. The next president must love America. The next president must embody unequivocally everything that is good about this country going back to its founding.

The next president must be the exact opposite of Barack Obama. He must be a man of high character and strong commitment to American values, because he will be facing problems and issues that no U.S. leader has had to face since the years leading up to World War II. In the late 1930s—only a few years before the Japanese attack on Pearl Harbor—our economy was still trying to recover from the Great Depression in the face of the policies of a big-government president, our military had been depleted, and our enemies were gathering strength and threatening war on many fronts.

While this was going on, our president and European leaders were hiding their heads in the sand, downplaying the threat from those enemies in much the same way Barack Obama is doing today. The Obama administration is concealing the seriousness of many problems that affect every one of us in our everyday lives.

Let me explain what I mean, by giving you information you won't hear from any other source:

- Obama's economic policies have caused such an enormous rise in the price of the necessities of life that his administration has removed energy and food prices from its financial analysis in order to avoid telling the American people just how bad inflation is.
- The U.S. Navy—our primary military force in the 21st century—currently has the smallest number of ships we have had since the end of World War I! In 1918, our navy was comprised of 774 ships, it's down to fewer than 300 today.[1] Bill Clinton decimated our military forces during the 1990s, and under Barack Obama the problem is getting worse.
- The gross domestic product of the United States is $15 trillion. The GDP of all the nations on earth together—including the United States—is about $60 trillion. The total debt owed by all the countries and banks and brokerages on the planet is estimated to be $600 trillion, ten times what all our economies combined produce in a year! It's the same debt that caused the economic crash of 2008—the one that sealed the deal for the election of the first Marxist president of the United States—only it's grown even bigger than it was then.

At best, Barack Obama will be doing nothing about these issues and the others I'm going to reveal to you here. At worst, he'll be actively trying to make these critical, potentially crippling problems worse.

In the face of this, the next president must be proud of America, willing to laugh at the lunatic fringe pushing the theory of global warming, courageous enough to take on the big bankers and the dictatorial enemies of freedom that are challenging us in every corner of the world.

The next president must do nothing less than restore America to its position as the supreme military and financial power in the world.

The American people have been kept in the dark about what I'm going to reveal to you in this book. They've been fed an endless stream of lies. They've had to endure out-and-out thievery by the government at their expense. They've had to stand by while the administration revealed its absolutely corrupt character—because the press kept its mouth shut.

Because of this, there is still the possibility that Barack Obama will be reelected.

You heard me.

Call it communism, Marxism, Leninism, socialism, collectivism, call it whatever you like. If we don't wake up to the reality of the situation, we face another four years of this administration, of the takeover of our country by agents of the very enemy that we've been fighting against for more than half a century.

We need to nominate and support the most electable Republican in the field, because an Obama victory in 2012 would doom this country.

I mean that literally.

We will very likely not survive as a free nation after four more years under Obama's rule.

It's never been more important that we vote a sitting president out of office. I'm going to explain why as I tell you what the next president of the United States must do to reverse the damage Obama has done. If the man chosen by Republicans to run against Barack Obama will follow the battle plan I outline here, explain to the American people that this is what he's going to do when he's elected, we will be able to save our nation from the most insidious and subversive administration in our history.

Returning American Foreign Policy to a Position of Strength

I've told the 10 million listeners to *The Savage Nation* radio broadcasts that the Obama foreign policy has been one of appeasing Islamist dictators and bowing to the Chinese and the Russians.

Let me start with Russia.

In 2009, Obama gave in to Russia when he abandoned the idea of installing U.S. missile defense sites in Poland and the Czech Republic that were meant to protect our European and Eastern European allies against Iranian nuclear missile strikes.[2] It was part of his Russian "reset." Russia rewarded Obama by threatening to aim its missiles at U.S. European-based missile defense systems and to withdraw from the new START nuclear weapons treaty it has with the United States.

You remember START, right? Obama promised to give up much of our nuclear weapons advantage over the Russians in the naïve interest of eliminating nuclear weapons from the world. The Obama strategy has been one of capitulation to despots around the world. His failure to stand up to the Russians puts the U.S. in greater and greater danger not just of losing our own weapons advantage but of putting our European allies at risk.

The next president of the United States must stand up to Russia, go ahead with the installation of missile defense sites in Poland and the Czech Republic while assuring former Soviet bloc countries that we will stand with them against possible Russian aggression.

China is not far behind Russia in challenging the United States.

Did you know that Chinese president Hu Jintao has directed the Chinese navy to prepare for military combat?

Hu's militant stance is based on China's claim that it "owns" the South China Sea and that other nations need to give up any claims to the oil and gas reserves there and to the right to use its shipping lanes without Chinese consent.

Do you know what our administration's response was?

The Pentagon called for "transparency" on China's part as the communist nation builds up its navy. Spokesman George Little talked about the "relationship" that the U.S. is "continuing to build with the Chinese military."[3]

Are you as appalled as I am at the lack of backbone displayed over and over again by our State and Defense departments in the face of China's ongoing military buildup?

The next president must proclaim publicly to the Chinese that the South China Sea represents international waters and that we will not tolerate any attempts on China's part to assert itself militarily there. The next president must actively seek the cooperation of countries bordering the South China Sea and of India and Japan in standing up to the Chinese communists on this issue.

We must also stand with India and its neighbors to deny China's growing presence in the Indian Ocean, where the Chinese are currently building naval bases. Improving our political and military alliance with India also strengthens us in the war against the expansion of radical Islam in the Middle East.

Have you taken a close look at what the Obama Middle East foreign policy has done to that region and to the global political balance since early 2011?

Obama supported the ouster of Hosni Mubarak during the spring of 2011.

In late November of that year, the first of a series of "popular" elections was held in Egypt. The Obama administration spent $200 million in support of the first Egyptian "democratic" election.[4]

The results? The radical Islamist Muslim Brotherhood, along with the conservative Salafist party, won 67 percent of the seats in the new Egyptian parliament, giving supporters of government based on Shariah law a strong majority.[5]

In Egypt, the military junta that took over the country has showed no signs of being ready to give up its power to the people. By November 2011, tens of thousands of Egyptians were again staging their own version of Occupy Tahrir Square, protesting the military government that their "revolution" brought into power only a few months before. As they did when they overthrew Mubarak, these Egyptian lovers of democracy sexually assaulted two more female journalists covering the new occupation, and the military killed dozens of demonstrators. The occupiers demanded that the military turn over power to the transition civilian government at the same time the interim civilian government leaders were handing in their resignations *en masse*.

Democratic rule has turned what naïve leftists around the world called the Arab Spring into the Arab Winter, a term I coined. I saw as though through a crystal ball what would follow Mubarak's fall. The Egyptian economy has been reduced to a trickle as tourism has dried up. Christians are an endangered species in Egypt. If the Egyptian military would stop intervening to maintain order and hand over that job to a civilian government, the mob rule that has become the norm in Cairo would look like a picnic. Utter chaos would follow, and the country would very likely sink into true anarchy.

The demonstrators don't understand how good they had it under deposed president Hosni Mubarak. And they don't understand that it's not going to get any better under the Muslim Brotherhood—if the military ever relinquishes its power—even after the long Egyptian cycle of elections is finished.

At the same time, Tunisia and Morocco both elected Islamist majorities in their parliamentary elections.[6]

Do you remember how the Obama administration celebrated when Libyan leader Moammar Ghadafi was killed? Do you remember what Hillary Clinton said?

Gloatingly she said, "We came, we saw, he died."

Did she oversee the execution of a head of state who begged for his life?

Are we the new Visigoths?

Ghadafi looks like an angel of mercy compared to the National Transitional Council that has taken over governing in Libya.

The independent militias that Ghadafi was able to keep in check have resurfaced. They promised to give up their weapons after Ghadafi was taken down, but they've reneged on those promises. Violent clashes and revenge killings have become the norm. I don't see any chance of a government that can centralize control over this tribal society and the warring militias forming anytime soon. The idea of a national army in a tribal state such as Libya is a joke without Ghadafi.

It doesn't get any more promising when I look beyond Libya and Egypt.

In November 2011, Great Britain severed banking ties with Iran, suspending all business relationships and bank transactions between British and Iranian banks. The Brits were worried that the financial sector was "being unknowingly used by Iranian banks for [nuclear] proliferation transactions."[7]

The Iranians responded by overrunning the British Embassy in Tehran, forcing Britain to withdraw its embassy personnel and to cease diplomatic relations with Iran. The embassy takeover was staged by the Iranian government in order to make it look like Iranian citizens supported the ruling regime.

Obama responded with this: "[F]or rioters essentially to be able to overrun the embassy and set it on fire is an indication that the Iranian government is not taking its international obligation seriously."

"Not taking its international obligation seriously?"

Since when has Iran ever taken its international obligation seriously?

Since when has Iran done anything except laugh behind Barack Obama's back as Obama neglected *his* international obligation to protect the U.S. and our allies when he failed to step in and stop a rogue terrorist nation from threatening the world?

Where the British actually instituted economic sanctions that will have a serious detrimental effect on Iran's economy, the U. S. has failed to do anything meaningful. In fact, when the Senate voted 100-0 on a bill to penalize foreign financial institutions that do business with Iran, Obama insisted that they "soften" their stance against the terrorist nation.[8] No mention by Barack Obama of the Iranian nuclear weapons program, no mention of their killing demonstrators who did rise up against the government in 2009, only a mealy-mouthed statement about "international obligations."

How do we change it?

The next president of the United States must follow the British lead and institute real economic sanctions on Iran . . . if it's even necessary. Because the first step the next president must take is to make it clear that we are Israel's ally and will stand with Israel in support of preemptive strikes against Iranian nuclear weapon production sites, if the Israelis haven't already taken them out.

The next president of the United States must reverse the support of Islamist governments and parties in the Middle East that has characterized the Obama presidency. The next president must abandon Obama's commitment to Islamist dictatorships and realign the U.S. with America's traditional allies, the oil-producing nations in the Middle East who understand and have made it clear to us that they want Iran's nuclear development stopped.

The next president must reestablish a military presence in the Middle East, possibly negotiating a return of our troops to Iraq, but minimally establishing and rebuilding our military forces in the region. Our mission must be based on defending our allies who recognize that Iran is out to do nothing less than take over the entire region and turn it into a new caliphate with designs on expanding throughout the world.

Above all, the next president must restore our relationship with our most important ally in the Middle East: Israel. In early December 2011, Defense Secretary Leon Panetta slammed Israel for its "isolationist" policies. He called on Israel to mend fences with Turkey and Egypt, two countries coming increasingly under Islamist rule, while he neglected to mention that the U.S. supported the overthrow of Egypt's Hosni Mubarak, one of the few allies supporting both Israel and the U.S. in the Middle East.[9]

In restoring U.S. ties with Israel, the next president must make it clear that Palestinians acknowledge the legitimacy of the nation of Israel and that that will be the basis of a Palestinian state. In addition, the next president must come out strongly against the anti-Semitism that has spread like a plague among liberals and leftists around the world, making it clear that U.S. support for Israel is uncompromising.

Restoring National Defense

If you look for this information in the newspapers and broadcasts of the Obama media acolytes, you won't find it: Under Barack Obama, defense spending has fallen to the smallest percentage of our gross domestic product since Bill Clinton's administration. Under Obama's rapidly shrinking defense budget, the fleet is set to be reduced even further. The 2012 defense budget of $513 billion represents only *3.4 percent of GDP.*[10] That's down from 5 percent only three years ago, and it's down from 6.2 percent during the Reagan administration. There's an additional $117.5 billion to fund the ongoing military actions in this budget, but Obama's cuts to the defense budget itself actually put our troops in danger of not having the logistical support they need in the field.[11]

The vote in the Senate to approve the defense spending bill was 93 to 7. Obama's response?

He threatened to veto it because there's a provision in the bill that says enemy combatants detained anywhere in the world, including in the United States, are subject to mandatory military custody.

Here's what Obama's veto means: *If you don't give me civilian trials for enemy combatants, I will not sign the defense spending bill.*

He's holding the U.S. military hostage to his desire to see terrorists on trial in American courtrooms. The threatened veto is part of Obama's ongoing effort to undermine U.S. national security.

Are you aware of the additional cuts Obama has in store for the military?

You remember the failed "select committee" that was supposed to find $1.2 trillion in budget cuts? Obama knew they would fail, so he set it up to automatically cut an additional $600 billion from the defense budget over the next ten years. That's another $60 billion a year. He did it through the automatic spending cuts that kicked in when the Select Budget Committee of six Democrats and six Republicans was unable to agree on how to cut $1.2 trillion from the federal budget.

Obama is threatening to veto any legislation that would prevent those automatic $60 billion annual additional defense cuts from kicking in, too.

Even leftist Defense Secretary Panetta, who backed some of the most devastating defense spending cuts in U.S. history during the Clinton administration, has said that the United States could become a "paper tiger" if further cuts are made.

Obama either doesn't understand or refuses to acknowledge that we are at war. We face continuing danger in Iraq and Afghanistan because of troop drawdowns and ongoing state-sponsored terrorism. Iran and North Korea are closer than we acknowledge to being able to deliver nuclear strikes through their missiles. Al Qaeda is still actively seeking nuclear and biological weapons capabilities. China and Russia are becoming more defiant and threatening in the face of the weakest U.S. president in our history.

Obama's already reduced defense spending and his threats to allow draconian defense cuts to kick in benefit our enemies. He's determined to cede our military superiority to Russia, China, even Iran, in order to hasten the formation of a new George Soros-defined world order, where national borders disappear and the rule of the international mob becomes a reality.

We are at war on many fronts, and the next president must increase defense spending as a percentage of GDP to wartime levels. That means returning defense spending, not including funding for military actions, to at least 5.2 percent of GDP—that's the 45-year average—while making sure that the drawdown in troops Obama has implemented is reversed and the buildup of new weapons, warships, and military aircraft is restarted.

Reversing the Obama Energy Policy

Here's something Barack Obama will never admit: Despite his efforts to cripple it, the oil industry is booming in the United States. The oil industry has accounted for more than 20 percent of the new jobs in the United States in the past seven years. Texas—which Obama is trying to give away to Mexico—has accounted for most of the rest of them.

What is the president's energy policy?

Obama and his bought-and-paid-for secretary of the Environmental Protection Agency, Lisa Jackson, have done everything in their power to keep America dependent on foreign oil. They've all but permanently banned offshore drilling in the United States, and they've managed to kill the Keystone XL pipeline so we can't buy and ship more oil from our ally, Canada, directly through to Gulf Coast refineries.

Obama's energy policy is also compromising America's electric grid and putting Americans at risk of not being able to afford electricity—if there's even enough to go around. After the freak October snowstorm in 2011 that knocked out power to millions in the New England states, some homeowners were without electricity for as long as a month before their power was restored. And Southern California has experienced rolling blackouts that are an omen of future electricity shortages throughout the Southwest.

Obama's assault on coal-fired electricity generation has meant that, even in the short span of three years, he's making sure that no new electric production facilities will be brought online. Obama has weakened the grid in many areas of the U.S. because he's bent on shutting down coal-fired electricity production, not in the name of making life better for humans but of making it better for plants and animals.

Obama's energy secretary, Steven Chu, has lied shamelessly to Congress and the American public about the threat of global warming. He's perpetuating the myth that our continued use of carbon-based fuels threatens the planet. He couldn't be further from the truth, as I'll show you in chapter 9. Like Obama, he favors windmills and electric cars and forcing energy prices high enough to make green technology viable.

Obama concurs: "I share the view of . . . most scientists in the world that climate change is a real problem and that human activity is contributing to it, and that we all have a responsibility to find ways to reduce our carbon emissions."

The fact is that he doesn't share the views of "most scientists"; he's parroting the party line about a phenomenon that has been in decline for the past decade and poses no real threat whatsoever—except that rising temperatures make things better for all forms of life on earth. He's spouting Stalinist science in the face of evidence he does his best to suppress that global warming is not a threat to our way of life. The real threat is a false belief that we have to protect ourselves from global warming. In fact, the path Obama is taking us down weakens our economy and our national security.

I've understood for a long time that the whole global warming debate should have ended in 2009 when it was revealed that so-called "scientists" who supported the idea that the earth was in a disastrous warming period had phonied up the numbers and corrupted climate science irrevocably. It didn't end there. In November 2011, more than 5,000 new e-mails sent back and forth among top climate "scientists" reveal that they're still fudging data, trying to intimidate those who disagree with them into silence. Despite the collapse of the evidence they themselves invented, they're still spouting the party line.[12]

Climate change is still the rationale behind this administration's push to close down the U.S. oil industry. Leftist Lisa Jackson still controls the future of the oil industry in the U.S. She can single-handedly dictate whether or not taking oil out of the ground goes ahead. Now she's even taken over the design of U.S. automobiles. A new diktat by Jackson's EPA will drive up the cost of automobiles and reduce their safety dramatically in an effort to meet artificially set CO_2 emission standards.[13]

What this administration hasn't been able to do is completely shut down the expansion of the oil industry. Across the country, more than 440,000 people are working in the oil industry. That's up from 200,000 less than a decade ago. North Dakota now has an unemployment rate under four percent because of the oil boom in that state. They're producing nearly half a million barrels of oil a day. In Pennsylvania, natural gas production from the Marcellus shale formation is turning around that state's economy. Nationwide, more than 200,000 people now work in the natural gas industry.[14]

It's new oil and gas technology—not green technology—that's driving this expansion.

Hydraulic fracturing—"fracking"—and horizontal drilling technology have made the extraction of oil and natural gas easier and safer than it has ever been. Fracking involves pumping a mixture of sand and fluid at high pressure into shale rock formations that contain oil and gas. This fractures the rock and frees up the oil and gas for recovery. It means that tens of billions of barrels of oil and virtually limitless supplies of natural gas in U.S. fields can now be taken out of the ground.

Horizontal drilling means that from a single well we can expand horizontally in several directions to remove the oil over a large field. We no longer need to drill many new wells to exploit the oil in a large field. This means that there is less environmental stress than ever before. And U.S. oil companies adhere to the strictest safety standards.

The new president must make sure that we can go ahead with the exploitation of our enormous oil and natural gas reserves. The next president must bring the EPA under control. He must make sure it relaxes its stranglehold on new offshore oil development, and he must reverse what Obama has just done and let the Keystone XL pipeline project go forward. He must combine that with allowing offshore drilling and the exploitation of known U.S. oil reserves.

This would make us energy independent before the current class of kindergartners graduates from high school.

Don't believe that?

If the United States announced tomorrow that it was allowing oil production to go forward in the Gulf of Mexico, off the California and Alaska coasts, and in other now undeveloped areas where we have the world's largest proven reserves,[15] the price of oil would stabilize—immediately and permanently. The markets would understand that, given our vast reserves of oil—enough to carry us through to the next century—a U.S. commitment to exploiting those reserves would stabilize the oil market.

When the next president gives the go-ahead to the Keystone pipeline, he'll insure that the oil will start flowing from Canada within a matter of years, not decades.

The positive effects would go way beyond what I've just mentioned.

Simply by relaxing the EPA stranglehold on oil and gas production, the next president would begin to economically isolate rogue regimes in the Middle East in a way no vague, weak-kneed "economic sanctions" of the kind Obama has talked about but never instituted could ever do.

If the next president goes ahead with what I'm calling for here, within as little as a decade, we would be able to supply not only all of our own oil and gas needs, but those of our European allies as well.

Opening up the U.S. oil industry would mean that our European allies would not have to remain dependent on Arab oil. If we'd relaxed these regulations even two or three years ago, the NATO nations would not have needed to oust Ghadafi in order to make sure they continued to receive Libyan oil.

I'm telling you that it is not out of the question to think that Obama's energy policy, because it has kept us beholden to Middle Eastern oil, was one of the primary causes of the chaos that's now unfolding as the Arab Winter wears on and Islamist forces advance their assumption of power.

The next president must change our energy policy to open up our oil industry as I've explained here. The policies and practices I'm calling for would solve many of our immediate economic problems.

Fixing the Global Economic Mess

I've explained to you that freeing up oil and natural gas production in this country would begin to turn our economy around almost immediately. That's because it would reintroduce real production back into the U.S. economy, real commodities on a scale large enough to dramatically increase U.S. GDP.

The broader problem that has infected the world economy—and the one that we must begin to address immediately—is that our economic and financial system is no longer tied to real commodities, services, and industrial production. Instead, the value of our currencies is at the mercy of fiscal and financial policies that have no intrinsic economic value.

Big banks and sovereign governments still carry hundreds of trillions of dollars' worth of the same worthless securities that caused the 2008 financial crisis on their balance sheets. They're trying to continue to finance their debt and cover up the huge holes in their balance sheets by issuing bonds that they buy from each other. This only exacerbates the problem, because they have to pay higher and higher rates of interest on these bonds in order to sell them, so they print more money in order to pay the interest on the debt.

The debt spiral will ultimately become a death spiral for the global economy.

The European economies that are about to collapse are in the shape they are largely because Europe has been, since the end of World War II, dominated by socialist governments. In order to finance the welfare states and the corruption that those socialist economies breed, they've had to rely more and more on debt, and the debt they try to sell to the rest of the world is essentially worthless.

More than half a century of European socialism is coming to the crisis point now, and the next president must act decisively to prevent a catastrophic global economic meltdown.

The European Union is in dire economic straits. At least three—and as many as seven—EU countries are on the verge of defaulting on their sovereign debt. In the near future, they're in danger of not being able to pay interest to the investors, the banks, brokerages, and other national governments that have bought their bonds. In order to prevent that from happening, the EU, along with the International Monetary Fund, is promoting a scheme for other countries to come together and pony up enough money to buy even more of these countries' debt—so they can avoid defaulting in the short term. The United States committed to helping fund this in late 2011.

The banks and federal agencies being called on to do this already hold much of this essentially worthless debt on their balance sheets, and adding more only makes it increasingly likely that these countries will downgrade the debt even further, making future default more likely.

There's not enough money in the world to prevent it.

Let me give you some concrete examples.

In late 2011, Italy—one of the countries about to default—managed to sell three-year sovereign bonds, but only after they were forced to pay nearly eight percent interest in order to get other financial institutions to buy them. That's an unsustainable rate. The European Central Bank continued to buy Italian and Spanish debt, but even their purchases were not at high enough levels to prevent what looks like a coming Eurozone financial crash. At the same time, the sale of sovereign bonds by Germany—the strongest and safest European economy—was a flop, with less than two-thirds of the offering purchased by investors. One analyst described the failed German bond auction as "a vote of no confidence against the entire euro zone." [16]

By the end of 2011, the euro was about to disappear as the currency of the European Union.

It goes without saying that the next president needs to get rid of Federal Reserve chairman Ben Bernanke and Treasury secretary Tim Geithner. What he needs to do next is more difficult.

He needs to extract the federal government from the clutches of the international financial elite, the banks and brokerages and sovereign governments that now dictate economic policy. The current administration is filled with former financial industry players, especially, as I'll show you in chapter 4, former Goldman Sachs executives. It's going to be very difficult to remove them from power, but it must be done.

The next president must stop the irresponsible printing of money that we've been engaging in in order to pay for our own and the world's borrowing. If possible, the next president needs to find a way to link our currency to a commodity like gold or to a group of commodities, so that the runaway inflation is slowed and the collapse of the debt bubble is avoided.

Cleaning Up Health Care

I've studied medicine and health and nutrition extensively, and I've written many books on the subjects. I understand what few medical doctors and no government bureaucrats do: that individuals—in almost every case—are responsible for their own health and the health of their children. Most of the diseases and conditions that kill us are the result of lousy choices we make about diet, exercise, and drug and alcohol use. Most of the maladies that are increasing at such alarming rates these days—from obesity to autoimmune disorders—are preventable. But not by the government telling you what you should and should not eat and determining what medicines you can and cannot take.

I won't repeat what I've said in *Trickle Up Poverty* about Obamacare. I will say this, though: Fixing the health-care mess is one of the easiest jobs facing the next president. If the Supreme Court doesn't shut down Obamacare by the summer of 2012, the next president needs to shut it down by executive order. Mitt Romney has said that's one of the first things he would do if he's elected.

Then the next president must do the following things, either through getting legislation passed or through executive orders:

- Repeal government requirements that dictate what treatments health insurance policies must cover, then offer smorgasbord insurance policies that let individuals tailor their health insurance to their own needs and control their own health-care spending. In other words, take it out of the hands of government.
- Legalize health insurance sales across state lines to let the market function to encourage competition that lowers prices.
- Make the cost of health insurance paid for by individuals tax deductible, as it is for corporations.
- Reform our bankrupt Medicare system to remove the threat of health-care rationing and encouraging market-driven choices for retirees. Gradually wean future retirees off this outmoded health insurance system through means testing and privatization of Medicare insurance.

It's the job of the next president to reverse the government takeover of one of our most important industries. It's the job of the next president to make getting the government out of the health insurance and health policy business his highest priority.

Securing Our Borders

Listeners to *The Savage Nation* know that one of my most important focuses is on our border security. They also know that the three people primarily responsible for securing our borders—Barack Obama, Eric Holder, and Janet Napolitano—are abdicating their constitutional duty to do so. In fact, they're not only failing to secure the borders, they're actively involved in just the opposite: They're opening them up so that illegal immigrants can cross them at will. They've sent assault weapons to Mexican drug cartels that have been used in hundreds of murders in Mexico, and at least three killings of U.S. citizens. And our own Drug Enforcement Agency has laundered drug cartel money!

The Justice Department and the Department of Homeland Security have quit trying to keep our borders secure, and they have attacked states like Arizona and Alabama that have passed their own illegal immigration and border security laws, taking them to court in order to stop them from defending themselves.

Did you know that under Barack Obama, the money we spend on border security has been cut in half? In 2008 we spent more than $1.3 billion to keep our borders safe, in 2011, that amount has been cut to $573 million. Janet Napolitano's Department of Homeland Security ended the Secure Border Initiative Network—the virtual border fence—because it did not meet "cost-effectiveness and viability standards." At the same time we're reducing the amount of money we spend on border security, the Obama administration—particularly Napolitano—keeps telling us the borders are more secure than they've ever been.[17]

It's not just that Obama, Attorney General Holder, and Napolitano are determined to eliminate our southern border with Mexico and give up Texas, Arizona, New Mexico, and California. They have lied to Congress and the American people about their intentions and their methods. In doing so, they've put our national security at risk. As I'll show you in chapter 10, they've compounded the felony by shutting out congressional investigators, withholding subpoenaed documents, lying under oath, and attempting to intimidate those called to testify in matters related to immigration and border safety.

It goes beyond securing our borders, though. For the very survival of our nation, the next president must make sure that people like Barack Obama, Eric Holder, and Janet Napolitano are driven out of our government and never again allowed to hold positions of power and influence of any kind in this country.

It's clear to me that Congress will not impeach Eric Holder and Janet Napolitano. The current crop of legislators lacks the will to throw them out of office. The American people must take the initiative at the ballot box. We must vote out the whole corrupt bunch of leftists that began a full frontal assault on America's borders, language, economy, and culture the day Obama took office.

In the wake of that, the next president needs to reinstitute the border fence and increase the manpower devoted to keeping our borders secured. Once that is accomplished, the next president needs to begin the process of restoring law and order to the immigration process. All criminal illegal aliens must be deported. The legislation that dictates that anyone born on American soil automatically becomes a U.S. citizen must be repealed. And the process of sorting out those in the country illegally and dealing with them in a way that follows the law must be begun.

Restoring the Constitution

As I explain in this book, Obama is a naked Marxist out to take over the United States as part of a larger plan to establish a collectivist world government. He's appointed leftist academics with no real-world experience to make every important decision, from foreign policy to economic strategy to energy to the military.

He's bowed down to foreign leaders around the world. He's determined to see the U.S. give in to the leftist and Islamist forces that are gaining power in the Middle East, Europe, and South America, where his foreign policy has focused on abandoning our former friends and making new friends of Islamist and communist dictators.

He's been able to do this because he's refused to follow the U.S. Constitution. The next president must clean house. He must get rid of the dictatorial cabinet secretaries and eliminate the appointment of czars who have commandeered the decision-making under this fascist presidency.

In other words, the next president must restore the Constitution to its rightful place as the foundation of our country's laws. He must make a new Declaration of Independence against the collectivists who have gained so much power in such a short time under Obama.

The Savage Revolution

Here's why the advice I'm giving is so important to the next president.

I worry that the Republicans don't truly understand what the Obama administration is doing to America. They submitted to the left-wing inquisitions that were passed off as "debates" during the Republican presidential primary, and they let themselves be attacked and destroyed by David Axelrod and the labor union cohorts who have formed mobs to perpetuate their corrupt ideology.

The Republican candidates sniped at each other, destroying their own credibility, as if the most important thing was which one of them got nominated, when what is really important is that we find a candidate who knows what is going on and can bring the message to the American people and do something about it when elected.

Part of the problem is this: As I've told my listeners over and over, there's no real difference between the Republicans and the Democrats. They've been trading eight-year presidential terms since 1980, with Republicans getting the presidency for two terms, then Democrats for two. At bottom, it doesn't matter which party you belong to if you're guilty of ignoring the will of the people and engaging in crony capitalism. By the way he handled the government, from the Fed to the Treasury to Defense, no American president except Ronald Reagan in the past 40 years has done what needed to be done to keep America on a sound military and financial footing.

Things changed in 2008 with the election of Barack Obama. As I explain to you in this book, we face a threat that goes far beyond any we've faced since before the fall of the Soviet Union. Barack Obama must be defeated in 2012 if our country is to survive as a free nation. Obama has made a pact with the devil and he's dragging the American people down with himself as he makes allies of the Islamist dictators in the Middle East and tries to ingratiate himself with the Russian, Chinese, and South American communist regimes.

It's time for the Savage Revolution.

I'm the only one explaining to you that Obama is in the process of taking our power to govern ourselves out of the hands of the American people and turning it over to a "world government" run by people who hate the U.S. and the freedom it represents.

Along with the Tea Party—which the Republican Party seems determined to damn into irrelevancy—I'm telling you that you've got to take back control of the country, through your vote and through the people and policies you support. You've got to take what I'm giving you in this book and translate it into action.

I explain to you that we have a chance to vote a naked Marxist president out of office. My work exposes the entire leftist mafia that controls the courts, the schools, and the economy.

The Savage Revolution will enable us to drive them out of power permanently. They'll still be here, but I am making sure they'll no longer be able to hide in the shadows, peddling their influence indirectly, affecting policy. They'll never again have the power to change the United States radically over a short time, as Obama has done.

This can only happen if the message of the Savage Revolution is carried by the Republican candidate.

The Savage Revolution will be built on a return to America's core values, a reestablishment of the Constitution as the basis of American law. It will mean that America once again asserts itself militarily and economically as the most powerful nation on earth, committed to standing against the tyrannical forces of radical Islam and the radical left. The Savage Revolution will mean developing policy for the good of the American people based on legitimate science and conservative economic principles. It will be characterized by enlightened regulation designed not to restrict American enterprise but to encourage it.

We must demand from our next president nothing less than a commitment to these values, policies, and actions and the restoration of American honor that has been so diminished under Barack Obama.

If Barack Obama is elected for another four-year term, he'll be president for life. He'll be the new Hugo Chavez. He'll do away with the two-term limit and win the 2016 election with 90 percent of the vote.

We have less than six months to make sure this doesn't happen.

CHAPTER 2

Tyranny of a Naked Marxist Presidency

Do you remember what happened on December 31, 2011, when most Americans weren't paying attention?

Let me remind you.

Barack Obama signed into law legislation that spelled out his power as president of the United States to detain any U.S. citizen indefinitely on the grounds that he or she might be a "terrorist." The National Defense Authorization Act (NDAA), which the president had earlier promised to veto, represents the single most egregious rollback of American civil liberties in our nation's history. Obama made the move under cover of America's New Year's Eve Party, while most Americans were more concerned with having a good time than with losing their freedom.

Even the normally supine ACLU and human rights leftists were stunned by the naked tyranny of the president's signing this bill into law.

Obama insisted that the reason he signed the bill was to guarantee funding for the military, to support our troops. In the process, he deprived U.S. citizens of the very freedom for which our military is fighting. In fact, despite his insistence to the contrary, it was the president himself who had fought for the right of the Commander-in-Chief to detain American citizens. Obama was outed by radical leftist legislator Carl Levin, who explained what the president was up to on the floor of the Senate. Levin made it clear that the White House had pushed for the law to be applied indiscriminately to American citizens.

In a signing statement issued along with the bill, the president said it was not his intention to use the law against American citizens, another executive statement that Obama clearly has no thought of honoring.

Why do I say that? Because the White House and the Justice Department have fought against bringing any cases involving indefinite detention of American citizens to trial for fear that the courts would overturn their right to abrogate citizens' constitutional guarantees of freedom from arbitrary arrest. In addition, the administration fought tooth and nail to insure that language barring the law from being applied to Americans was not included.[1]

There is widespread agreement in the legal profession that NDAA violates the U.S. Constitution. But the issue goes much further than simply agreement among lawyers.

Barack Obama has, in three short years, made rapid progress toward bringing the United States under his despotic rule. While America remains blissfully unaware of what he's doing, this president has accumulated power and authority over every aspect of our lives. This last affront must be the final subversive act we allow Barack Obama to commit. It must be the last time we allow him to strip our country of the guarantees provided in a Constitution Barack Obama not only does not believe in but continually ignores as he takes us closer and closer to his goal of state control of the country.

The president's latest assault on liberty brings up an important question: Do average Americans have anything to fear if Barack Obama and his cronies retain power after the 2012 elections?

In the following pages, I present the evidence that you haven't seen anywhere else. That evidence will convince you that you need to be very worried, and not simply about what this president might do to grab more power for himself and take away more of our freedoms if he is reelected. As the absence of outrage in response to his signing of NDAA indicates, most of you haven't been paying enough attention to what he's already done.

America's future hangs in the balance like a loose tooth. While there may not be time to save the tooth, We the People can stop the infection of corruption and power-madness that is going on all around us.

As I've told you, the president has in so many cases simply taken the law into his own hands, disregarding the Constitution, the rule of law itself. It doesn't stop with domestic policy. In addition to the affront to our civil rights that NDAA constitutes, Obama has long been trampling on the Constitution in dealing with U.S. citizens who are enemies of the administration. He hasn't stopped with simply imprisoning them: He's gone so far as to kill them.

While most of us rightly celebrate taking Anwar al-Awlaki—the American-born terrorist who masterminded al Qaeda operations from his base in Yemen—off the street, we need to pause and ask some serious questions.

I'll explain why.

First, al-Awlaki's murder was the precursor to the president's blanket approval—through his signing of the NDAA legislation—of rebuking American citizens' rights. The assassination was conducted under the authority of a secret memo produced by the Justice Department that justified the killing of an American citizen without affording him his right to the due process of law.

The über-liberal *Atlantic* magazine joined the Council on American-Islamic Relations (CAIR) and the American Civil Liberties Union (ACLU) in condemning the assassination. Normally I wouldn't give their opinions the time of day. They're anti-American and their pronouncements should generally be ignored. In this case, though, the *Atlantic* uncovered evidence that gives legitimacy to their being worried: Eric Holder's Anti-Justice Department developed a secret memo that authorized the terrorist's murder—without any other administration lawyers questioning the legality of the action.

The document justifies the assassination of al-Awlaki despite the fact that there is an executive order that bans assassinations and there are protections in the Bill of Rights against such murders. The memo justifies killing an American citizen if he can't be captured. It was apparently written specifically so the administration could take out al-Awlaki.[2]

It's not just that Obama has killed an American citizen without affording him his due process rights: He's done it *in secret*, refusing to have the DOJ release any documentation of the legal reasoning behind the strike!

Do you understand how Obama's sense of justice works? He and his attorney general want to try foreign national terrorists in the United States, giving them the same rights as U.S. citizens. Yet without giving him so much as a hearing, they assassinate a U.S. citizen accused of the same atrocities as foreign terrorists, without giving al-Awlaki the benefit of a trial.

Jury trials for foreign terrorists, assassination for American citizens.

If Obama can clandestinely kill an American citizen—however deserving he is of being killed—without even making clear the grounds on which he had him taken out, what's to stop Obama from using the same premise to take out other Americans with whom he disagrees?

The Most Corrupt Administration in Our History

I've told you over and over on my radio show that Barack Obama is presiding over the most corrupt administration in this country's history. He's bringing Chicago-style politics to Washington. Let me give you an idea of some of the things he's done that you won't hear from the mainstream media.

Did you hear that the Obama administration has sold the rights to processing our election results to a private company? That this critical component of free elections—the transparent tabulation of votes—will now be handled not by individual precincts but by a company over which we will have little control?

SOE, a U.S. elections results processing company which has handled processing election results for over 500 American jurisdictions, has typically provided prompt reporting of precinct-level results. However, SOE was recently acquired by SCYTL, a vote-processing company based in Spain, and this compromises the process enormously. Instead of producing election results trackable at the precinct level, votes will be transmitted to a central server, where they will be "counted." The problem is that once the votes are merged, it will be impossible to go back and check their integrity at the local level.

It is very likely that this is the final step in Barack Obama's corruption of the voting process. It has the promise of enabling him and his cohorts to control the outcomes of federal elections with no accountability. On top of that, it's one more step toward the formation of a "global government," one of the aims of George Soros, whose puppet currently occupies the White House.

Did you know that both Barack and Michelle Obama "voluntarily" gave up their law licenses and are not permitted to practice law in the United States? It's the same sort of voluntary surrender that Bill Clinton made after he lied under oath.

Are you aware of just how extensive the corruption was when Obama turned stimulus money into a slush fund for his political fundraisers and corporate cronies, pushing "green energy" projects like Solyndra that line the pockets of his crony capitalist contributors at the expense of American taxpayers?

You've heard about the "Fast and Furious" project, which put thousands of guns in the hands of the Mexican drug cartels and resulted in the deaths of two American border agents, but do you know that the president and his Attorney General, Eric Holder, dreamed up Fast and Furious, that both were in on it from the start?

How much do you know about the "czars" Obama has appointed to do the work that the president's cabinet is supposed to do? You may have heard about a few of them, but unless you've been listening to my show, you probably don't know the extent of this end run around your elected officials. Obama relies on dozens of these unvetted leftist cronies, and they have unchecked influence over economic and policy issues. They report only to him, and the administration has told us nothing about who they really are.

Have you heard about the Occupy Wall Street movement? I'm sure you have. You've probably run into the dirty little punks who are living on the streets of our cities in "Obamavilles"—the contemporary equivalent of the Hoovervilles of the Great Depression. But unless you read further in this book, you won't know the extent of the influence of American communists and influential political leftists like George Soros on the Puppet in the White House who's masquerading as president.

I'm sure you were watching television in the spring of 2011 when Obama loosed the union thugs that are the "muscle" he relies on to create chaos in the streets as part of his takeover strategy. But how much do you know about their role in intimidating voters in the state of Wisconsin and the bankers Obama wanted to intimidate and make an example of?

And are you aware that Obama deprived stockholders of their investments in General Motors and Chrysler, instead handing the companies over to the same union thugs who are the modern-day storm troopers enforcing his will on the American people?

In the middle of this utter corruption, *Newsweek* magazine's Jonathan Alter published an article titled "The Obama Miracle, a White House Free of Scandal." Here's what Alter said about the Solyndra scandal, the one in which Obama gave money to his benefactors who had invested in a green energy scam that was never intended to be successful: Solyndra was, as far as Alter was concerned, "not criminal or even unethical on the part of the administration." The stimulus money—our stimulus money—underwent "a rigorous process supervised by Vice President Joe Biden," who "prevented widespread fraud and abuse." According to Alter, the only reason project Fast and Furious is a problem is that "Representative Darrell Issa, the Republican from California . . . calls the administration 'corrupt' without offering any evidence."

Alter, himself a committed leftist, cites ultra-liberal political scientist Brendan Nyhan, who has declared "Obama set a record earlier this month for most days without a scandal of any president since 1977."[3]

Do you realize what these nitwits are saying?

The last scandal-free president, according to these two, was Jimmy Carter.

Do you understand what we're up against when utter fabrications like this are allowed to see the light of day?

The First Socialist Lady

Don't believe we've got a petty dictator in the White House?

Barack and Michelle celebrated the halfway mark of the schoolboy president's junior year in office with a trip to Martha's Vineyard. The vacation, in which Michelle and Barack took separate flights, cost taxpayers tens of thousands of unnecessary dollars. Michelle ordered up a separate military jet for herself and her entourage, arriving about four hours earlier than her husband, who flew in on Air Force One along with the family dog, Bo.

This wasn't the first time the First Lady and the First Kids had traveled separately from the president for no good reason. In December 2010, she left early for the family's Hawaii vacation at an extra cost of $100,000 to taxpayers.[4]

Her Martha's Vineyard trip capped Michelle Obama's two-and-a-half-year, $10 million vacation spending spree that included five-star hotels, massages, and top-shelf wines and liquors. Calculations are that Michelle spent at least 42 days on vacation from the fall of 2010 to the fall of 2011. She's been described as a "getaway junkie."[5]

My listeners are on to Michelle Obama. Here's what one of them wrote:

The one event that really did it for me was Mrs. Obama insisting on taking a separate flight . . . and all the cost that entails . . . to Martha's Vineyard, which apparently only netted her an additional 4 hours more than if she waited and went on Air Force One with the president. If that isn't telling the American public "Screw you, I'll do what I want," I don't know what is. Of course she's aware of the supposed austerity push going on, aren't we all? Apparently Obama has no control whatsoever over her actions. I've always been curious just what she did as a hospital administrator that warranted her being given a $300,000 plus salary (by way of Valerie Jarrett) just before the election in '08. The Chicago way, don't you know.

The First Lady is in good company. You remember Imelda Marcos, wife of Philippines dictator Ferdinand Marcos. She set the standard for lavish spending by a dictator's spouse by owning 3,000 pairs of shoes.

First Lady Michelle Obama is our own Imelda Marcos, keeping alive the tradition that dictators' wives should spend the money collected from their citizens lavishly on things that demonstrate their superiority to common people but, like our foreign policy, have nothing to do with our national interest. It cost U.S. taxpayers more than half a million dollars to send Michelle and her daughters—who were listed as "senior staff" on the trip manifest—along with the first lady's mother, niece, nephew, hairstylist, and makeup artists on an African safari that was billed as "government business."[6]

Did you know that the U.S. is being sued over this secret "family outing"? The organization Judicial Watch—which has already demanded to know how Nancy Pelosi spent more than $100,000 on her own birthday bash—now wants detailed financial records and passenger manifests for the First Socialist Lady's trip to Botswana. The government declined to hand over the information, so the organization has had to sue for it.[7]

About this trip, another of my listeners had this to say: "How exactly does a safari on a private preserve help poor, starving Africans? So they can see what a fat American African looks like? Is that the deal? So they can see a big jet filled with nouveau-riche poser criminals acting like potentates? Was that the plan?"

The Obamas finished off 2011 with their most expensive vacation junket yet: a Hawaii trip estimated to cost nearly $4 million, a good chunk of which will be paid for by U.S. taxpayers. The president and the First Lady traveled separately, with Michelle racking up roughly $100,000 in personnel, travel, and security expenses while the president, flying in on Air Force One at a cost of $181,757 per hour, will have spent more than $3 million by the time he completes the nine-hour round trip.[8]

Beyond the irresponsible use of taxpayer money, there's the first family's attitude toward our country. Michelle Obama revealed with five words just how great her and her husband's lack of respect—indeed, their utter contempt—for America is. While attending a 9/11 memorial ceremony at the World Trade Center in which the American flag was folded in the traditional manner, the First Socialist Lady rolled her eyes and said to her husband, "All this over a flag?" to which Barack Obama grinned and nodded his head in agreement.[9]

That sarcastic contempt for everything American, everything truly democratic, carries over into her political views. She's in favor of having schools feed our children, because the kids spend half their waking hours in schools and schools shape "the choices they're going to make for the rest of their lives." She's in agreement with Marx and Engels, who favored eliminating the role of the family in raising children. Michelle argues that the government "should provide clear, actionable guidance to . . . families on how to increase physical activity, improve nutrition, and reduce screen time."[10]

In other words, her supposed campaign against obesity is nothing more than an excuse to have the state take over the raising of America's children.

The First Socialist Lady is only another in a long line of wives and par-amours of dictatorial leaders, living the high life while she and her husband do everything in their power to keep their subjects from doing the same.

Are you starting to understand what's been happening in this rogue administration?

Why Do I Call It *Trickle Down Tyranny?*

Anyone who knows me or listens to my radio program will tell you that I choose my words carefully. I do exhaustive research and consider all the evidence. So when I tell you that Obama is a tyrant with a relentless desire to undermine our country, I mean what I say and I know what I'm talking about.

And when you read *Trickle Down Tyranny*, you will, too.

You're not going to find the evidence of Obama's tyranny in the pages of your local newspaper or on a nightly newscast. Fox News won't touch it. Even your favorite blogger doesn't have this. You have to follow me into the dusty stacks of government archives, peer into the pages of overseas news-papers that don't worship at Obama's shrine, and go where members of the Government-Media Complex fear to tread.

One of the things the network news and print journalists have refused to recognize is that we're heading for a dictatorship. No one understands that when you put the government in the hands of someone who's been brainwashed into believing the leftist tripe that passes for course content in American's colleges and universities you're submitting yourself to a tyrant.

Tyranny starts at the top. The people don't cause tyranny, their rulers do.

The last time the country was faced with having to rid itself of a ruler like the one currently occupying the White House was in 1776. As part of freeing themselves from the rule of the British king, America's Founding Fathers compiled a long list of what makes a tyrant and what constitutes tyrannical abuses.

As I go through that list in 2011, I find that Obama has committed many of the abuses catalogued in the Declaration of Independence.

The only thing he hasn't done is hire Hessian mercenaries to attack Americans in their homes and on their farms. In place of the foreign mercenaries, the Obama government now uses tax collectors, environmental regulators, zoning inspectors, and every conceivable bureaucrat to invade our homes, farms, factories, and businesses!

In other words, Barack Obama fits our founders' definition of a tyrant. Here is what they wrote:

But when a long train of abuses and usurpations, pursuing invariably the same object evinces a design to reduce them under absolute despotism, it is their right, it is their duty, to throw off such government, and to provide new guards for their future security. . . . The history of the present King of Great Britain is a history of repeated injuries and usurpations, all having in direct object the establishment of an absolute tyranny over these states. To prove this, let facts be submitted to a candid world.

Like King George III, Obama, too, has become a despot, trampling on the rights of individuals and state governments.

Let's look at some of the words of 1776 and the facts of 2011.

He has forbidden his governors to pass laws of immediate and pressing importance, unless suspended in their operation till his assent should be obtained; and when so suspended, he has utterly neglected to attend to them.

Through his Attorney General, Eric Holder—who I consider one of the most, if not the most corrupt person in the most corrupt administration in American history—Obama has sued the state of Arizona in order to stop the implementation of that state's immigration law, SB 1070. The law that Arizonans wrote in order to protect their border from further infiltration by Mexican drug gangs. The law that was necessary because the United States government under Obama was intentionally failing to carry out one of its most important constitutional duties—keeping our borders safe—in order to advance its subversive agenda.

He has called together legislative bodies at places unusual, uncomfortable, and distant from the depository of their public re-

cords, for the sole purpose of fatiguing them into compliance
with his measures.

The Obama equivalent of the King's crime is this: He has created more
boards and special commissions than any other president since FDR. Virtually all of their meetings are *secret* and the press and the public are *barred*.
These secretive task forces—which created the health-care bill, climate-
control legislation, and many other measures—do not release records of
their meetings or in many cases even reveal who is attending them to the
public. Yet out of them come far-reaching new laws that give the government the right to control the lives of more than 300 million Americans.

He has dissolved representative houses repeatedly, for opposing
with manly firmness his invasions on the rights of the people.

I told you in the Prologue how Obama is bypassing Congress and using
executive orders to implement his policies. In another move, Obama has
managed to engineer the *de facto* dissolution of Congress through his violation of the War Powers Act. The law requires the president to receive congressional approval for any military actions overseas that last more than 60
days. Harold Koh, the Obama appointee who is the State Department's top
legal advisor, tried to get around the law and the Constitution by contend-
ing that the then-90-day-old mission over Libya was limited to air strikes
and therefore, as ABC News has noted, "does not constitute hostilities as
described in the War Powers Act and therefore would not require Congressional authorization." [11]
Obama's authorizing his Energy Department to make taxpayer-
guaranteed loans to solar energy companies that are virtually guaranteed
to fail also bypasses congressional oversight. That's because Congress might
not be in such a hurry to have you and me guarantee that Obama's campaign contributors don't lose money on these nearly bankrupt companies.
The words of the Founders of our nation ring true today: *In every stage of
these oppressions we have petitioned for redress in the most humble terms: Our
repeated petitions have been answered only by repeated injury. A prince, whose*

character is thus marked by every act which may define a tyrant, is unfit to be the ruler of a free people.

As I will show you in this book, Obama is a tyrant in every sense of the word. He's defied the will of the people as they expressed it in the resounding defeats they administered to Democrats in the 2010 midterm elections. He has ignored the pleas of the American people, shrugged off the rulings of state and federal court judges—going so far as to mock and insult the Supreme Court of the United States in his 2011 State of Union speech—and defied the attempts of Congress to shut down his illegal maneuvers.

The Obama Tyranny

In *Trickle Up Poverty* I explained to you that the first two years of the Obama presidency amounted to nothing less than a leftist takeover of the United States. Early in his first term, with Democratic majorities in both the House and Senate, Obama pushed through sweeping legislation to take over the banks and automakers, seized control of health care, and took command of the student-loan market, eliminating private lenders.

When he couldn't get even his own Democratic-controlled Congress to do his bidding, Obama found ways to bypass the lawmakers and judges and secretly implemented his subversive agenda via the rapidly expanding federal bureaucracy.

Let me take you through what you'll read in *Trickle Down Tyranny* so you'll get a sense of just how dangerous this subversive administration has become in only three short years.

In chapter 3, "Tyranny of the Government-Media Complex," you'll discover how Obama continues to subvert our free speech rights.

One of the techniques is the use of intimidation. A CBS reporter who has reported on the Department of Justice's Fast and Furious scandal since it became public early in 2011 explained how the Obama administration tries to bully legitimate reporters into submission: "Well the DOJ woman [Tracy Schmaler] was . . . yelling at me. The guy from the White House on Friday night literally screamed at me and cussed at me."[12]

When a *Boston Herald* reporter sought to cover Obama's trip to Boston, White House press secretary Jay Carney pointedly excluded him. He wanted only the loyalist White House press corps to cover the president when he visits American voters. When Vice President Joe Biden was speaking at a $500-per-plate fund-raiser in Florida, *Orlando Sentinel* reporter Scott Powers was roughed up and locked in a closet for several hours by presidential goons. The president himself has attacked Fox News and tried to exclude it from covering public events where the president is present. Obama has also put together a "war room" to use Facebook, Twitter, and other social media to send out his propaganda without the checks and balances of a free press.

The so-called press has become the *Pravda* of the 21st century under Barack Obama.

In chapter 4, "Tyranny of the Treasurer," I address one of the most important issues of our time: U.S. debt. In that chapter, you'll read about how Obama has refused to rein in spending, growing the national debt from $9 trillion in 2009, when he took office, to more than $15 trillion in 2012. It took America from 1789 to 2008 to rack up $9 trillion in federal debt—it took Obama less than three years to increase that 200-year debt load by nearly 70 percent. He has refused to reform runaway so-called entitlements or cut payments that go to his supporters and donors. In Obama's first three years, the national debt climbed from slightly more than 40 percent of our gross domestic product to more than 100 percent of GDP. Rather than cut spending, Obama spent the summer of 2011 fighting with Congress to increase federal debt by another $2.5 trillion.

But Obama can't stop spending even on frivolous items. He's nearly doubled the number of limousines used by government officials. In the last year of the George W. Bush administration, the federal government owned 238 limousines, according to the General Services Administration. By 2010, two years into Obama's reign, the federal fleet had swelled to 412 limousines. Those limousines aren't cheap. The Government Accountability Office (GAO) said that the Obama administration spent $1.9 billion to buy and operate limousines in 2009 alone. So let's crunch the numbers. With a fleet of 412 limousines and a budget of $1.9 billion, that means that each vehicle costs the taxpayers just over $4.6 million. Even including the cost of the driver, security detail, higher gas prices, comprehensive insurance, maintenance, and parking fees, it is hard to see how the federal government can possibly spend $4.6 million per car. But they do, according to a March 2011 GAO report.[13]

Who knows? Maybe they park it next to the $400-dollar hammer and the $2,000-dollar toilet seat? Leslie Paige of Citizens Against Government Waste, a Washington, D.C.-based nonprofit, wasn't amused. "It's one more reason why there is so much cynicism in the public about what goes on in Washington."[14]

But beyond that, I'll explain to you in chapter 4 how Goldman Sachs executives have essentially taken over the finances of the United States, and how they've instituted policies that have endangered the world economy by creating new collateralized debt that amounts to more than ten times the world GDP.

Chapter 5, "Tyranny of Obama's Radical Accomplices," explains how the White House has become labor union headquarters under Barack Obama. Former Service Employees International Union (SEIU) capo Andy Stern and AFL-CIO president Richard Trumka signed the White House Guest Register dozens of times as they cemented labor and public employee unions' hold on this utterly corrupt administration. Labor unions have stood behind the Occupy Wall Street demonstrators and the mobs that destroyed public property in Wisconsin when they tried to shut down the state government after it had trimmed public employee unions' bargaining rights.

To pile corruption on top of corruption, Obama has also drafted an executive order that says businesses seeking federal contracts must reveal who they made political contributions to, despite the fact that this is clearly a violation of our constitutional rights under law and the fact that federal contracts are required to go to the lowest bidder.

It doesn't stop there.

When the Ford Motor Company—the one U.S. automaker that did not accept bailout money—aired a commercial in which a Ford owner made it clear that he wouldn't buy a car from a company that had accepted bailout money, Obama bullied Ford into stopping the broadcast of the commercial.

As you'll read in chapter 5, the leftist fondness for mob rule goes back to one of the most dangerous of radicals, Saul Alinsky. Obama was mentored in the art of "community organizing" by Alinsky disciples, and his administration is littered with leftists who adhere to the methods put forth in Alinsky's book *Rules for Radicals*.

In chapter 6, "Tyranny of Obama's Corporate Cronies," I'll explain to you how Obama has given over his administration to leftist thugs that he's appointed as czars, bypassing the U.S. Constitution with no oversight or approval from America's elected officials. The salaries of the czars are not reported to the public and they carry out duties only known to the president himself.

These unvetted lieutenants extend the administration's illegal control over large segments of the U.S. economy, from automakers (Ed Montgomery, Director of Recovery for Auto Communities and Workers) and oil producers (former Energy Czar Carol Browner) to college retirement funds (former Fannie Mae CEO Herb Allison) and marital relations (Lynn Rosenthal, White House Advisor on Violence Against Women). His Pay Czar, Ken Feinberg, was appointed to dictate what Americans can earn in the financial services industry and, eventually, in all industries.

In chapter 7, "Tyranny of the Egghead Wars," I'll take you through how the eggheads in the Obama administration—none of whom has a bit of experience in the real world—have hijacked America's foreign policy and put it at the mercy of what's known as the "Responsibility to Protect" doctrine, a political philosophy developed by disciples of one of the most dangerous men in the world, George Soros.

This doctrine lies at the foundation of the so-called Arab Spring—which has become the Arab Winter after the ouster of American allies in countries throughout the Middle East and their ongoing replacement with Islamic extremist governments. It's the realization of the vision of American decline and the rise of dictatorships favored by Soros and his anti-Semitic, anti-American confederates to encircle and destroy the Jewish state of Israel as the first step in eliminating democracies around the world.

In chapter 8, "Tyranny of Treating Our Friends Like Enemies and Our Enemies Like Friends," I take you on a tour of the world and explain to you how the Obama foreign policy has caused us to abandon the political allies who we've stood with since World War II and replace them with enemies of freedom, "allies" who are committed to the downfall of the United States.

The president's foreign policy has been laughed at and ridiculed by the countries he has sought as our new allies, from Russia to Iran. Obama has encouraged the ouster of longtime American allies in the Middle East who have managed for the past 40 years to maintain a semblance of stability in that region, protecting Israel's right to exist and promoting U.S. interests in the region. Now Obama turns his back as our new allies Iran and Syria kill their own citizens—who are truly committed to the overthrow of these two governments—while he supports the formation of new Islamist governments in countries that have gotten rid of leaders sympathetic to the United States.

It's the most perverse, anti-American, anti-Israel policy of any president since the end of World War II, and it's going to lead to a new caliphate in the Middle East, headed by a nuclear-armed Iran.

Chapter 9, "Tyranny of Green Energy," talks about another of the most important components of the Obama takeover. After you've read this chapter, you'll understand that Obama cares nothing about the environment or how energy is produced or how much pollution it creates. The sole purpose of his energy policy is to take power out of the hands of the people and transfer it to the U.S. government.

After acting with "determined disregard" (in the words of a federal judge) in summarily shutting down oil drilling in the Gulf of Mexico during his inept handling of the 2010 Gulf oil spill, the Obama administration was found to be in contempt of court when it defied a federal judge's order to remove his illegal ban. By that time, the number of deepwater rigs in the Gulf had shrunk to eight, from the more than 30 that had been in operation before the spill and before Obama had imposed his drilling moratorium. Tens of thousands of jobs disappeared when those rigs were forced to be abandoned by the destructive adolescent in the White House.

After the House and Senate had declined to pass cap-and-trade legislation, Obama decided to use the Environmental Protection Agency (EPA) to implement policies that he himself said were guaranteed to make electricity prices "skyrocket." In doing so, he moved closer to his goal of taking over the U.S. energy industry by first crippling it, then commandeering it in much the same way he did the auto industry.

He followed this up with the "green energy" scandals. First, there was the $535 million taxpayer-guaranteed loan to political cronies at the failed solar shell corporation Solyndra, followed by a $737 million loan to another phony solar company—this one with direct ties to Solyndra and to Nancy Pelosi—only two weeks after the FBI had raided Solyndra.

Green energy has become a synonym for corrupt cronyism in the Obama administration. It has nothing to do with energy policy and everything to do with shutting down America's energy production in the name of keeping political donors happy.

Chapter 10, "Tyranny of the Anti-Justice Department," deals with the person I think is one of the most corrupt ever to occupy the office of Attorney General. Eric Holder effectively legalized limited amnesty for illegal aliens with the stroke of his pen when he signed an executive order designed to placate racist La Raza and other antidemocratic Hispanic pressure groups. He decided unilaterally not to enforce the Defense of Marriage Act, which defines marriages as "only a legal union between one man and one woman as husband and wife," despite the fact that the law was passed by an 85–14 majority in the Senate and 342–67 in the House. In doing so, he'd taken the right to define marriage away from the Congress and the states, where it belongs, and effectively put it in the hands of Attorney General Eric Holder. Holder is the federal official who described the Black Panthers as "my people" when he got fed up with questions about how he handled a voting rights case involving that radical political group.

The Transportation Security Administration (TSA) is allowed to use powerful and dangerous X-ray scanners to peer through the clothes of passengers. Hundreds of virtually nude scans of young women and famous actors were almost immediately posted on the Internet—the very thing Obama said wouldn't happen. Not even infants or the elderly were exempt from the invasive scans.

If you didn't want to be exposed to dangerous levels of radiation, you had to let a TSA agent grope you instead. Obama refuses to learn from Israel, which without invasive scans or pat-downs still manages to prevent terrorists from flying. But did you know that Eric Holder's Justice Department has moved to ban flights out of Texas if that state makes TSA groping illegal?

At the time, I was the only one who explained to you what this move was really about: It represents one more way Obama is softening up the American people, getting them used to having their individual rights violated and their freedoms taken away. We now passively accept the TSA's often perverse intrusions into our most personal space, and the Justice Department's threats if groping is banned.

These are not minor bureaucratic moves, but dangerous, large-scale takeovers that violate the Constitution and the rule of law.

In chapter 11, "Crushing Obama's Cadre Before They Crush Us," I'll give you a summary of the Obama administration's inexorable progress toward turning this country into a dictatorship not unlike that Hitler managed to put in place in Nazi Germany. The parallels are so disturbing that they call to mind the words of Milton Mayer, who wrote about how Hitler's takeover of Germany was accomplished over time, much as Obama's takeover of the U.S. has been.

Obama's Tyranny Brings Loss of Freedom

The president is staging nothing less than an assault on our freedoms, including the freedom to worship as we choose. His focus is on undermining Christianity and Judaism as they become increasingly hostile to the freedom of religion guaranteed in the First Amendment. At the Houston National Cemetery, a graveyard for veterans that is run by the U.S. Department of Veterans Affairs, director Arlene Ocasio has banned the use of the words *God* and *Jesus Christ* in funeral services, set a limit of 20 minutes for any military funeral, refused to have taps played at the funerals of homeless veterans, and required that all prayers be submitted in advance for government approval. This Janet Napolitano look-alike is a lifetime low-level administrator who managed to work herself up to a position of authority where she could impose her atheist will on our military veterans' families.

The American people are increasingly making their feelings about this administration known in public opinion polls. In an October 2011 Gallup poll, the president's disapproval numbers were 52 percent, equal to those of the Russian despot Vladimir Putin. In other words, Americans are as fearful of the tyrannical nature of the Obama administration as they are of that of the Russian tyrant. The news website *WorldNetDaily* publishes the results of an ongoing "Freedom Index" survey conducted by the polling organization Wenzel Strategies. The poll represents, in the words of company CEO Fritz Wenzel, Americans' "deeply held public perceptions about their freedoms and the government's impact on them." The Freedom Index "dropped off a cliff" as Obama and his Democratic allies in Congress pushed through the government takeover of the health insurance industry. Americans particularly fear the loss of privacy as the government gains control of their medical records and intimate personal details that they reveal only in whispers to their doctors in a quiet exam room with the door firmly closed.

I'm not surprised at all.

I've been telling you since before the earliest days of this administration that we're giving away our freedoms and our rights to this Usurper-in-Chief and that until we take back control from the thieves Obama has appointed to steal them from us, we're facing an existential challenge to this nation unlike anything we've experienced since 1776 or under FDR, who, like Obama, favored big government and strong-arm tactics against ordinary citizens.

Americans are right to worry.

They're beginning to understand that Obama is a tyrant, out to take everything from them and put it in the hands of the federal government.

Other polls show that Americans are increasingly anxious about losing their economic rights in the Obama years. Seventy-five percent of Americans surveyed favor an amendment to the Constitution forcing the Congress to balance the federal budget every year, according to a May 2011 Wenzel Strategies poll. Americans fear that Washington is spending them into the poorhouse, bankrupting the country and loading down future generations with backbreaking debt for generations to come. Even Democrats agree, with 59 percent responding in the May 2011 telephone survey that the federal government's value for money is "only fair" or "poor."

Why are Americans worried about coming under the rule of a tyrant?

Maybe it's because under Obama we're seeing things like these: A single 68-year-old man in a nursing home will now be forced to buy a costly insurance plan that covers pregnancy and skateboarding injuries.

Or maybe it's because the president has further destroyed American culture by spending hundreds of millions of dollars to promote "bilingual education" programs that discourage immigrants from learning English and learning about our country's unique history and culture.

Or it could be that the world's largest economy now has the world's largest debt, and that debt is larger than the total debt of the nine next largest economies in the world combined?

No wonder we're hearing demands like this one from William Gheen, president of Americans for Legal Immigration PAC, an organization representing more than 40,000 Americans who want to see tougher enforcement at the border: "By arming drug and human smugglers with assault weapons that have been used to kill American and Mexican citizens and police forces, and by ordering amnesty for illegal aliens, which has been rejected by both Congress and the American public more than eight times, Obama has committed a form of treason against the United States and must be removed from office by Congress."

Gheen is not alone in calling for impeachment. *WorldNetDaily* founder Joseph Farrah recently wrote: "Never before in the history of the United States has an occupant of the White House displayed less concern for the Constitution and the rule of law than Barack Obama. It's about time somebody said it: It's time to impeach Obama." [15] [16]

Larry Klayman, former Justice Department prosecutor and founder of Judicial Watch, puts it succinctly: "[F]rom what has already occurred, the offenses of the mullah-in-chief are already so compelling as to warrant immediate impeachment and conviction for his high crimes and misdemeanors." [17]

The Final Chapter: A Tyrant's Consolidation of Power

What we're being subjected to is "consolidation," a tactic employed by leftists and other fascistic political groups once they gain a foothold in order to bring more and more of the decision-making power normally granted to the states and private enterprise—where those powers should lawfully reside—under the control of the federal government through unchecked regulation and behind-closed-doors policy decisions. In doing this, Obama—classic narcissist that he is—is remaking this country in his own image, turning it into something that reflects the belief system he was brainwashed into accepting by the leftist mentors of his youth.

By usurping the legislative and judiciary functions of our federal government, and by siding with cohort groups like leftist labor unions in wage and benefits disputes, Obama has done what no other president since FDR had even thought of doing: He's installed himself as Chief Executive, Summary Legislator, and Judge and Jury in every critical decision originally granted to the states and individuals through our representatives by the Constitution.

We've got a radically unqualified man in the highest office in the land. But Barack Obama is much more than simply out of his depth: There is no doubt in my mind that he's consciously involved in one of the most dangerous attempts our nation has faced to rapidly steal our freedoms and replace them with a tyrannical government that controls every element of our lives.

This is why I call it *Trickle Down Tyranny*. The president is effectively dismissing our constitutional rights and sending us up the road to becoming a third-world dictatorship as he hands down judgments and issues policies diametrically opposed to the beliefs of an overwhelming majority of American citizens and enshrined in the United States Constitution!

It's my purpose to explore and expose this existential threat to our democracy—to democracies around the world—and to develop the arguments and strategies we need to put into practice in order to take back our country.

I wrote this book because America needs another Paul Revere, a truth teller to wake and warn Americans that their freedom is slipping away. We need a leader who will draft another Declaration of Independence, this one against the tyrant in the White House.

The American people will know such a leader when they see him . . . or her. They'll rally behind that leader as Americans rallied behind George Washington more than two centuries ago in order to oust a dictatorial king.

This book is a call to action in the night of a suffering country.

If you want more evidence of Obama's tyranny, you need to read this book. And then tell your friends and neighbors. By each one of us becoming individual Paul Reveres riding through our towns and villages armed with the truth that shall set us free, we will save this country.

CHAPTER 3

Tyranny of the
Government-Media Complex

I am no friend of Rupert Murdoch.

And the dislike is mutual. The media baron, who owns the Fox News Channel, *The Wall Street Journal*, and many other newspapers and television outlets in the United States, the United Kingdom, Australia, and Asia, doesn't like me. Fox News doesn't cover my speeches or follow up on the scoops on my radio show. I'm never invited to appear on its cable news network. I could drop dead and Murdoch wouldn't care. And I don't owe Fox or Murdoch anything, either.

Like I said, I'm not a friend of Murdoch. And it is not my job to write a legal defense of that strange media mogul. I just call them as I see them, without fear or favor.

Yet, in the interest of "fair and balanced" assessment of the news, I think it's time to put certain aspects of the media war into a fresh perspective—one that you haven't heard anywhere before because no one else dares to tell you the truth like I do—and show you how the media elite and the federal government are conspiring to take away your rights. I am talking about the tyranny of the "Government-Media Complex."

In his last days in office, President Eisenhower warned the nation about the gathering strength of the "military-industrial complex." As a career U.S.

Army officer and former Supreme Allied Commander in Europe during World War II, his warning was sincere and powerful.

In March 1998, at the famous Commonwealth Club of San Francisco, I warned the public about the growing power of the Government-Media Complex. As a radio host with an audience of more than 10 million listeners and author of many bestselling books, I know the media inside and out. Like Eisenhower, I know what I am talking about. And I have the guts to say it. In my speech, I demonstrated that the elite media—the television networks, the major newspapers, the big magazines, the radio stations and the wire services—were aligned with the governing class of the country.

The press is supposed to "comfort the afflicted and afflict the comfortable," but now they are cozy cronies of the government.

They are supposed to be watchdogs, but now they are lapdogs.

They are supposed to be Rottweilers, but now they are Shih-Tzus.

And they carry out their servile duties in two ways: highlighting the stories that the government crowd wants pushed and ignoring the stories that hold the government accountable.

Here's what I told the Commonwealth Club on March 12, 1998. As you can see, all of it is still accurate today:

> It is censorship by default as well. What do I mean by censorship by default? Where the commercial interests of the media moguls are so intertwined with government policy as to create an overly friendly image of government officials and their policies. It may not be as clear as a close conspiracy to bias the news, but it results in the same form of censorship of dissent.
>
> Now this bias is not limited to the left. It is largely a product of left-wing bias when it comes to certain social issues such as Affirmative Action and "gay" rights. But the right also biases the news when it wants to shape fiscal issues to its benefit. Example, Rupert Murdoch and the China scandal a while back.
>
> I first began my file on the Government-Media Complex several years ago. I noticed an alarming bias, and I knew this could sink the ship of truth. Surely other administrations have had their

friends in high media places. Still, there were many voices and many views of dissent that found their way into the national media. But now we have a growing media blackout of some serious crimes and misdemeanors, all unsolved to the satisfaction of those with critical faculties of reason. . . .

Each and every issue as reflected in the old-line news media—that is, the TV network news and the establishment newspapers and magazines—is a parallel reflection of official Clinton policy.

Listen to this carefully if you will. Tell me if you agree. Hoover Institution historian Robert Conquest said that in the former Soviet Union the press was totally under the control of the state. All editors were members of the Communist Party. Here in the United States of America a frighteningly imbalanced Washington press corps exists. Eighty-nine percent of these apparatchiks of the DNC voted for Bill Clinton in 1992!

Let me repeat, "The media is needed by the public to be and remain a thorn in the side of the government in order to keep the government relatively honest. But when the media instead becomes a thorn in the side of the skeptical private citizen, the media then becomes an arm of the government."

Is this not worrisome? "Beware the Government-Media Complex."

And today, the Government-Media Complex is more dangerous and more powerful than ever. In a moment, I will show you irrefutable evidence that the so-called mainstream media wouldn't dare report.

But first let's take a tour of the Government-Media Complex.

The New *Pravda*

When I was kid in the 1950s, journalists and government officials didn't go to school together, marry each other, or go into business together. Journalism was a working-class trade and there were few Ivy League graduates in the newsroom. For that matter, most journalists didn't even go to college, much less major in journalism. Journalists started out on a beat. They covered city hall or the police department and worked their way up from there, with hard-nosed editors questioning everything that they wrote.

That experience, which often lasted for decades before they were promoted, shaped them into a skeptical jury of ordinary citizens with extraordinary access and information. By the time they made it to Washington, they weren't impressed by senators, ambassadors, or even presidents. They had covered politicians for decades and knew the breed well. They were suspicious, even hostile. They knew that they were the "people's intelligence service" and had to be tough to earn their readers' trust. And they enjoyed catching the politicians trying to pull a fast one.

Today, journalists go to Ivy League schools and *start* in Washington, D.C. They attend the same schools they send their kids to. More to the point, they attend the same schools as government officials. They socialize with politicians and brag about them coming to dinner parties at their homes.

Let me give you an example of something you've never heard in the mainstream media: Tammy Haddad. She's had a varied career as a producer at CNN, Fox News, NBC, CBS, and MSNBC. She's the head of Haddad Media, a company that develops Internet and event programming. She's also the former executive producer of Chris Matthew's MSNBC show.[1] Haddad throws a garden party every year, right before the White House Correspondents Association Dinner, and she eagerly tweets the arrival of every politician to her front yard.

The Correspondents Association Dinner itself is such a love-in between the president and the press that Obama joked at the 2009 dinner, "All of you voted for me."[2] He won huge applause for that line because the nearly 2,000 members of the working press in the room knew it to be true. Surveys repeatedly show that some 91 percent of the broadcast and print media vote for Democrats.[3]

And the press now marries government officials. Consider the case of NBC on-air reporter Andrea Mitchell, who in 1997 married then-chairman of the United States Federal Reserve Alan Greenspan. Or the case of President Obama's ambassador to the United Nations, Susan Rice. She is married to an ABC News executive. ABC's Christiane Amanpour is married to a former Bill Clinton State Department spokesman.

Other journalists treat the wall between the press and the people they cover as a revolving door. Chris Matthews was a top assistant to House Speaker Tip O'Neill. After leaving that job, he wrote a book and became a bureau chief for the *San Francisco Chronicle*, penning a weekly column for that paper. Next, he got his own television show, called *Hardball*.[4]

And he's far from the only one going through the revolving door between government and media. Dana Perino, a former Bush White House spokeswoman, now anchors a show on Fox News Channel called *The Five*. Bill Clinton's former ambassador to Morocco, Marc Ginsberg, is now a columnist for the Huffington Post and a producer of Arabic-language television.[5] Tony Blankley, a former spokesman for House Speaker Newt Gingrich, is now a syndicated columnist and Fox News talking head.[6] Mary Matalin, previously the Republican Party spokeswoman, is now heading a division of Simon & Schuster, the book publishing outfit.[7] Her husband, James Carville, a key Clinton campaign operative in the 1990s, writes a newspaper column and regularly appears on cable news programs.[8]

And so on.

If you've wondered why you hear the same thing from every news outlet in the country, it's because "big media" has such a streak of ideological conformity. It is the same people pushing the same agenda.

It's why you hear all about global warming, but never about the "climate-gate" scandal.

It's why you hear all about the need for gun control, but never about the government's botched "Fast and Furious" operation to sell guns to murderous Mexican cartels.

It's why you hear all about the benefits of diversity, but never learn about the crimes committed by illegals.

It's why the media defend the big budgets of the federal and state governments but never mention the fact that the average federal worker ($123,000 annual salary) earns more than twice as much as the average citizen (at $52,000 per year, the median U.S. household income).[9]

It's why you hear all about the poor, but never about welfare cheats or "homeless" criminals.

That's why you hear all about soldiers and CIA agents allegedly violating some terrorist's civil rights, but never about the brave soldiers or intrepid CIA men who save lives.

The Government-Media Complex is just one big happy family, with journalists marrying or going into business with government officials or becoming government officials themselves. Media and government are now one interchangeable class and, like all aristocracies, they seek power by taking away your rights.

Here's how they do it.

The Soros-Murdoch Media War

There is a colossal battle going on right now between two of the most powerful media figures on this planet: Rupert Murdoch and George Soros.

Here's what you need to know about Soros: He owns Barack Obama lock, stock, and barrel. Barack Obama is George Soros's factotum—it's not the other way around—and Soros is using Obama and the entire U.S. press to fight Murdoch and to bring him down. And right now, Soros is winning.

Soros's attack on Murdoch began shortly after Fox News's Glenn Beck ran a series of attacks on Soros. You might remember that early in 2011 there was a constant stream of anti-Soros material on Fox. The network emphasized the idea that the tax-exempt status of one Soros organization, Media Matters—which is dedicated to pushing Soros's fascistic, left-wing, anti-America agenda—should be canceled.

I know what I'm talking about. Media Matters staged a five-year attack on me. I think they're behind getting me banned in Britain. Media Matters took extracts from my radio broadcasts, made over many years, and edited them in order to slander me. They put me through a year of legal hell defending myself against their lies. And now the British government has affirmed that I'm still banned in Britain. Why? Because they say I'm unable to prove that I "did not commit . . . unacceptable behavior." By those standards, we could accuse anyone of anything we wanted to, then convict him because he couldn't prove he didn't do it. It's nothing less than another indictment of the false and slanderous attacks of leftist governments on those of us who are telling the truth about what's happening as the world descends into chaos.

But to get back to the Soros-Fox conflict: In fact, it was Beck's continual hammering of Soros and his connections to leftist causes that led to Media Matters getting Fox News to dismiss Glenn Beck. Beck's show is no longer on the air at Fox.

There's no question that Media Matters' tax-exempt status should be revoked. Media Matters is an overtly political organization and has no business being a tax-exempt corporation. Shortly after this confrontation, Soros's attack on Murdoch began in earnest, and it's playing out in the on-going battle between the two media Goliaths.

It looks to me like Soros is winning.

The one threat to the Government-Media Complex is the truth: the facts and evidence that don't fit their official storyline. That's what you find on radio shows and websites like mine. And sometimes, even Fox News Channel departs from the politically correct view of the news. Billionaire media mogul that he is, George Soros can't stand the fact that Murdoch, another billionaire media mogul, sometimes stands in his way and gives the American people news that threatens Soros's aims.

Before he was taken off the air, Fox's Glenn Beck—like me—was always talking about Soros's complex plans to subvert our democracy. That made Soros hate Fox News even more.

When Fox News went after MediaMatters.org—the website funded by Soros' millions—Soros's response, as he told Poltico.com, was to donate $1 million to Media Matters "to hold Fox News accountable for the false and misleading information they so often broadcast."[10]

And just so you don't mistake what Soros's views are, here are the words of the man himself, speaking on CNN's *Fareed Zakaria GPS*: "Fox News makes a habit—it has imported the methods of George Orwell, you know, newspeak, where you can tell the people falsehoods and deceive them. . . . They succeeded in—in Germany, where the Weimar Republic collapsed and you had a Nazi regime follow it. So this is a very, very dangerous way of deceiving people. And I would like people to be aware that they are being deceived."[11]

Get it?

He thinks Fox News is trying to bring a Nazi-like regime to power in America.

Sounds to me like a case of projection—he is imputing to Fox what he intends to do himself: bring about a radical and tyrannical change in America's government.

Soros is a left-wing Murdoch. He helps fund National Public Radio and has ties to at least 30 other major media organizations charged with policing media bias throughout the industry.[12]

In other words, Soros is trying to gain control over the media watchdogs.

While this was going on, Media Matters' head David Brock declared "war"—yes, that's the word Media Matters used—on Fox News, telling Politico.com that he was launching an all-out campaign of "guerilla war and sabotage" against Fox. He added that he was using a "strategy of containment" to keep Fox scoops from being picked up by other networks or newspapers. He announced that he has a paid staff of 90 people arrayed in a war room near the White House.[13]

Why Fox? Brock says it is the "nerve center" of the conservative movement.

This is the same strategy that Soros uses in Europe—where they call it *cordon sanitaire*—to keep popular political parties that question Muslim immigration policies or the welfare state from forming a parliamentary majority and enacting policies supported by the majority of voters. It has successfully prevented the Vlaams Belang, the largest political party in Belgium, from ever forming a government and electing a prime minister, simply because all of the other political parties refuse to go into coalition with it, claiming it has a "racist" anti-Muslim platform.[14]

In reality, Soros's minions want to keep the other center-right parties from campaigning or governing on the immigration and cultural issues. The same Soros-inspired *cordon sanitaire* is in effect in France, Germany, Austria, the Netherlands and Sweden. It's one of the ways Soros helps keep the illegal migration of Muslims into Europe flowing: by threatening to defame any party or person that would shut it off.

Back to Media Matters and its "war" on Fox News and Soros's attempts to implement a U.S. version of the *cordon sanitaire* on the news organization.

Whatever you think of Fox, America is better off with more news channels presenting different points of view. That is why the Founders gave us the First Amendment's free speech rights. But Media Matters wants to isolate Fox and then destroy it. That means your right to read, hear, and see alternative points of view will be taken away. You will be left with only the Government-Media Complex version of the news.

In other words, you will be left in the dark.

Fox fought back. Media Matters is a tax-exempt 501(c)3 "educational" organization. In order to qualify for tax-exempt status, the organization has to prove it has a valid educational purpose. Destroying the nation's most popular cable news network is not exactly like teaching little kids about American history. It is pure political partisanship, and that is not supposed to be tax-exempt. So Fox's news anchors and reporters began producing stories that questioned whether Media Matters should enjoy its special status with the IRS.[15]

Why do Soros and the establishment media hate Murdoch so much?

What really bothers Soros and the media is that Murdoch offers a profitable alternative that beats them in ratings and readers. Murdoch broke the unions in the United Kingdom and weakened them in the United States, and his papers' continuous stream of scoops makes the left look lazy and clueless.

Soros has already destroyed Conrad Black's media empire—Black at one time owned the Telegraph newspaper group, the *Chicago Sun-Times*, and Canada's *National Post*, among other outlets[16]—the only other conservative alternative to the left's media monopoly. He wants to do the same thing to Murdoch, so that the public will be denied its right to an alternative free press. He only cares about the First Amendment when it helps him. Soros and his leftist allies really want a one-party press, like Syria or Iran.

When Media Matters was attacked by Fox, Soros fought back. Within weeks of Fox's coverage of Media Matters, the British press discovered a "scandal." Murdoch-owned newspapers in the United Kingdom were paying private detectives and listening in on phone calls to get personal information about government officials, professional athletes, movie stars, and even the royal family.[17] These are not exactly military secrets, but it made the establishment very angry. They don't like to be embarrassed. But these gossipy items thrilled readers, who bought Murdoch's newspapers to read the latest. So, all around the world, newspapers are dying while in Britain Murdoch found a way to make them thrive.

Of course, the rival media barons, stoked by Soros's front groups, said this was a scandal. And it was. The British detectives appeared to have violated British laws and violated the privacy of hundreds of people over the years.

But the mainstream media in the United States didn't tell you the full story. Remember, they compete with Murdoch, too, and they love Soros as much as they loathe Murdoch.

Again, I'm not a fan of Rupert Murdoch, but I'm here to tell you the truth, no matter who it helps or hurts.

Here are some things that the Government-Media Complex doesn't want you to know.

There is a real public benefit to at least some of what Murdoch's papers did. Murdoch's *News of the World* newspaper, while Rebekah Brooks was its editor, violated the privacy rights of convicted pedophiles to reveal to the public that dozens and dozens of them were out of prison and living in neighborhoods with young families with small children.[18]

If you had a child and lived next door to a pedophile, do you think the government should protect his privacy rights over the safety of your children?

Neither did most Britons.

When they learned about the dangerous criminal living in their midst, they were angry and demanded changes. That uproar led to the passage of "Sarah's Law" in the British parliament, giving the public right to access government databases to learn the home addresses of convicted pedophiles. Soros's minions screamed that the pedophiles' privacy rights were violated and that they had paid their debt to society, but it didn't matter. Most of these child-molesting felons were forced to move away from neighborhoods packed with the very children that they once preyed upon.

Another thing that the Government-Media Complex won't tell you is that most British papers routinely hire detectives and trample on privacy rights. What Murdoch's papers did was routine in Britain. Indeed, I found a 2006 report called "What price privacy now?"—developed for Parliament by the United Kingdom's Information Commissioner's Office—that showed that some 305 journalists from 21 different publications committed some 11,000 violations of the Data Protection Act, Britain's privacy law.

The vast majority of violations were by non-Murdoch publications and the number-one offender was a rival of Murdoch. After a lengthy investigation that year, the Crown Prosecution Service declined to press any charges against any journalists.[19]

Hacking into the cell phones of Britain's royal family in 2007 is a different story. That year, the royal correspondent for a Murdoch-owned paper and a private detective were briefly jailed for listening in on the royal family's voice-mail messages. The story uncovered by the crime was hugely important in Britain though. It revealed that the heir to the throne, Prince William, had injured his knee and that his ability to walk again was in doubt. The royal family hadn't told the public and the subjects were eager to learn about the health of their future king.[20]

Arguably, that story was also in the public interest. Importantly, prosecutors said there was no evidence that any of Murdoch's editors were involved and no one else was charged. The scandal did not go all the way to the top of the organization.

But Soros needed to bloody Murdoch, his rival. So his allies hyped up old charges from the 2004–2006 period, the time frame already examined by the police, prosecutors, and Parliament. That's another key detail that the Government-Media Complex won't admit: These charges are old and probably won't stick.

In the hoopla, Parliament asked Rupert Murdoch and his son James to testify. They did. Murdoch called it the "most humbling day of my life" and spent millions buying full-page advertisements in other people's newspapers to apologize to the British public. But the two Murdochs never admitted to any wrongdoing.

In fact, the old buzzard got a round of sympathy when a crazed activist interrupted the parliamentary hearing and tried to put a shaving-cream pie in his face. Murdoch's wife, Wendi Deng, slapped him back on the forehead, driving him away from her husband. Murdoch and his wife got good press out of that incident.[21]

Indeed, Soros's bid for revenge may backfire. British prosecutors announced that they were also investigating the *Daily Star*,[22] a paper not owned by Murdoch. Again, something the Government-Media Complex doesn't want you to know.

And CNN may be the next unintended victim snared in the Soros trap. Piers Morgan filled the chair of CNN's legendary softball-question king, Larry King. Now he may also be filling a seat in jail cell. That's because he wrote a little-noticed book called *The Insider: The Private Diaries of a Scandalous Decade*, in 2005. In that book, he writes: "Apparently if you don't change the standard security code that every phone comes with, then anyone can call your number and, if you don't answer, tap in the standard four digit code to hear all your messages." Next, he commented: "I'll change mine just in case, but it makes me wonder how many public figures and celebrities are aware of this little trick."

Now he's being investigated for phone hacking.[23]

Did I mention that he was the editor of left-liberal newspaper that competed with Murdoch? It seems that Murdoch's rivals are up to their necks in the same "crimes." Again, you didn't see that in any of the major newspapers because the Government-Media Complex doesn't want to cloud your mind with facts that conflict with its official line.

So Soros switched tactics again. One of his minions wrote a piece in the *Daily Beast*, which is owned by Tina Brown and the Harmon family that owns *Newsweek* magazine. If you are an audiophile, you know the Harmon family name from its Harmon Kardon line of speakers and other audio equipment. Brown and the Harmons are fully paid-up members of the media elite and the Government-Media Complex. The *Daily Beast* article called for the prosecution of Murdoch on suspicion that his British newspapers' payment of British detectives violated America's "Foreign Corrupt Practices Act." That law is supposed to stop American companies from bribing foreign officials to get contracts from foreign governments.

There were no Americans involved and no British government contracts were involved. So prosecuting any News International executive under that law would be a real stretch. But that didn't stop the *Daily Beast* author from salivating, "Imagine an Eric Holder–appointed advisor supervising Fox News." [24]

And that's the real agenda here. Bring Fox News under the control of the Government-Media Complex and make them toe the official line.

And, sure enough, the FBI immediately announced it was opening an investigation into Murdoch's media holdings to see if—you guessed it—any of his executives (or Murdoch himself) violated the Foreign Corrupt Practices Act.[25]

And guess whose jurisdiction the FBI falls under. That's right: Eric Holder.

That's how the complex works. The government and media work together hand in glove.

The Government-Media Complex Always Wins

Did you know that Barack Obama—the head of the Government-Media Complex—has ordered his minions at the newspapers and television outlets under his jurisdiction to "minimize references to al Qaeda?"

That's right.

The president doesn't want you to hear the name al Qaeda associated with the 9/11 terrorist attacks on our country. He'd rather his scribes in the Government-Media Complex "honor all victims of terrorism, in every nation . . . whether in New York or Nairobi, Bali or Belfast, Mumbai or Manila, or Lahore or London."

Note the alliteration, the fancy turn of phrase.

Note the idiocy of this punk president's pronouncement.

Instead of honoring America's dead on this most solemn occasion, Obama wants to "present a positive, forward-looking narrative."

Until November 2011, I didn't know of another case in which the president has intervened so directly in an attempt to control the press as he did with these words.

I'm talking about the day the president took over all broadcast networks, overriding all broadcast programming.

On November 9, 2011, the president tested a project that's been in the background for half a century but had never been tested: the Emergency Broadcast System. It was ostensibly designed "to provide the President of the United States with an expeditious method of communicating with the American public in the event of war, threat of war, or grave national crisis."[26]

I don't buy that.

To me, it looks like the president is rehearsing for the time—in the very near future—when he declares a state of emergency in the United States and takes over public broadcasting on a more permanent basis.

He's setting it up with the Occupy Wall Street protests in cities across the nation. He's setting it up by having his union thugs create chaos like they have in Wisconsin, where they took over and trashed the state capitol. He's setting it up with the class warfare he's trying to incite between the haves and the have-nots.

He's ready to initiate another takeover of our rights as citizens—the right to a free press—in the same way his TSA has done. That organization continually violates our Fourth Amendment rights against unreasonable searches and seizures when we pass through the airport screening process.

Obama is getting ready to extend the unconstitutional even further.

He's just making sure the equipment works so that when the day comes, he'll be able to step in and shut down our communications networks—in the name of maintaining order.

My question is this: Does Barack Obama think we're so naïve as to believe that this is just a test? It's a test all right. It's a test by the government to make sure they can shut down communications and deny the American people the information they need in case of political emergency.

Do you trust a single word from any member of the Obama administration?

Have you ever heard anything but lies and disinformation out of these America-haters?

What this tells me is that the Obama regime foresees an imminent national emergency that will necessitate that it takes over the avenues of communications so that it can control, even more than it does through the Government-Media Complex, everything we hear and see.

The Rise of Government Regulation

As if the fact that the media have become nothing more than a venue for the presentation of a leftist agenda wasn't egregious enough, Obama is out to extend government control over news outlets even further.

I'm talking about government regulation of the media.

I don't favor government investigations of news outlets. It is too close to tyranny, to government control of the free press. But if we are going to be fair about it, why not look into a major American newspaper that has its own phone-hacking scandal and that has repeatedly violated our national security by printing classified material? Indeed, the paper I'm referring to puts our soldiers and spies at risk while we are fighting the war on terror.

That newspaper is the *New York Times*. In fact, any "crime" committed by Murdoch's operation has actually been committed over and over by the *New York Crimes*.

Let's go where the mainstream media refuse to tread.

In 1996, a Florida couple, John and Alice Martin, used a radio scanner to hack into a cell phone conversation between Representative John Boehner and House Speaker Newt Gingrich. They made tapes of the illegally obtained conversation and gave them to Democratic congressman James McDermott, a member of the House Ethics Committee. Coincidentally, the Committee just happened to be investigating Gingrich at the time. McDermott immediately handed them over to the *Crimes*. The page-one story in the *Crimes* embarrassed the Republicans, especially Gingrich.[27]

That was the idea, to embarrass Republicans.

It gives you a sense of what the word *ethics* means to a Democrat.

And while the hackers were later forced by a court to pay a $1 million fine, the *Times* got off scot-free.

Or consider all the classified information that the *Times* obtained illegally and then made public—to the detriment of our war effort. In December 2005, the paper published classified information that revealed that the U.S. government was listening in on calls made on cell phones that al Qaeda operatives had bought in Switzerland.[28] Within 24 hours, all those al Qaeda phones went dead.

Six months later, Uncle Sam's classified program to track the money that funds terror attacks was reported by the *Times*.[29] It was a global effort, involving dozens of European and Arab countries. Most feared the political consequences of helping the U.S. stop al Qaeda—any seeming support for President Bush would hurt them with their people. So as soon as the news broke, America's allies ran for the hills. All al Qaeda bank transfers disappeared, and European and Arab governments stopped helping us track the money.

The *Times* also reported that the National Security Agency was listening in to the international calls of terror suspects in the United States. That's right: Someone with the private cell phone number of Osama bin Laden was dialing it from inside the United States.

Jack Bauer would listen in and so would you.

Don't you think we should know what the terrorists inside our country were planning?

That same year, the *Times* published the details of CIA secret prisons.[30] Allied governments stopped cooperating and all prisoners had to be shifted to Gitmo, effectively ending all interrogations. Without those secret prisons and severe methods, all usable intelligence from al Qaeda prisoners stopped.

In 2009, the *Times* published secret Bush-era Justice Department memos on interrogation and said the CIA were "torturers." Some CIA officials retired while others now face criminal prosecution. They will win, but their colleagues won't be pushing the envelope to uncover terror plots.

Why should they risk their careers?

Jack Bauer should just go home.

In fact, the Espionage Act—section 793(e) of U.S. Code Title 18—does apply to the editors of the *New York Times*, and they could be prosecuted. Those editors could face up to ten years in prison or tens of thousands of dollars in fines. The *New York Times* appears to have specifically violated Section 793(e) of Title 18 of the U.S. Code. The relevant section clearly applies:

> Whoever having unauthorized possession of, access to, or control over any document, writing, code book, signal book, sketch, photograph, photographic negative, blueprint, plan, map, model, instrument, appliance, or note relating to the national defense, or information relating to the national defense which information the possessor has reason to believe could be used to the injury of the United States or to the advantage of any foreign nation, willfully communicates, delivers, transmits or causes to be communicated, delivered, or transmitted, or attempts to communicate, deliver, trans-

mit or cause to be communicated, delivered, or transmitted the same
to any person not entitled to receive it, or willfully retains the same
and fails to deliver it to the officer or employee of the United States
entitled to receive it.

The case is clear-cut enough. The newspaper is not cleared to receive
classified documents, yet they received them and even published them. The
paper also failed to promptly turn over the secret papers to the proper government officials and even used those illegally obtained documents "to the
injury of the United States."

Case closed.

Think the First Amendment's free-speech protections mean the law
doesn't apply to the *New York Times*?

Think again.

The United States Supreme Court considered that very question shortly
after the law was adopted during World War I. In *Schenck v. United States*,
the Supreme Court held that those who were convicted under the law had
no shield under the First Amendment. The Constitution was designed to
protect people who wanted to help America's enemies during a shooting war.

And what about the Pentagon Papers case?

Daniel Ellsberg and Anthony Russo were working for the Pentagon in
1970 when they decided to secretly photocopy thousands of pages of classified intelligence about the Vietnam War. Justice Department prosecutors
won a grand jury indictment against the pair in 1971, for allegedly violating
the Espionage Act. They appealed, ultimately landing before the U.S. Supreme Court. In the *New York Times Co. v. the United States,* the high court
held that government could not stop the newspaper from publishing the illegally gotten documents, but—and this is the key part of the finding—the
government could still constitutionally try, convict, and punish editors for
violating the Espionage Act if they did publish the classified material. So the
Supreme Court said that no one has a constitutional right to publish secret
documents.

Why did Ellsberg and Russo later get let off?

The official story is that the judge declared a mistrial. In reality, the
Government-Media Complex protects its own.

So while the *New York Times* is champing at the bit for the government to go after Murdoch, the "paper of record" has done far worse things itself.

But don't expect the FBI, which opened an investigation into Murdoch, to expand its inquiry into the activities of the *New York Times*. Why? Because the Government-Media Complex always wins in the end.

The Tyranny of Blame: Norway's Christian Killer

Let me give you a vivid example of the Government-Media Complex at work.

Remember Anders Breivik, the Norwegian nut who killed more than 90 people in a bombing of Norway's prime minister's office in Oslo and the grisly gunning down of scores of teenagers on an island retreat?

First, let's remember some hard truths. Not all Muslims are terrorists and not all terrorists are Muslims.

Think about it. If all Muslims were terrorists there would be 1.1 billion Muslims at war with everyone else on earth. Muslims may hate you, but they're not terrorists, just as you may hate others, but you're not a terrorist. Don't tell me you've got a halo on your head just because you have a cross and go to church.

Look within your own heart and you will know I am right. I expect many of you are angry with me right now. You think I'm totally wrong. But after you have a chance to reflect on this incident, you'll say, "Savage, you were right."

You need to understand the danger of certain ideas and where they can lead you.

Now, given all this, Breivik wrote about al Qaeda with admiration, and that explains why this madman killed his own kind, hunted down young Norwegian children like rabbits. His mass murder was the most dastardly, Nazi-like act I've seen in my time. It's the same as I've seen here in America with guys in the black raincoats. I am talking about the Columbine High School killings in Littleton, Colorado, in 1999, when two youths who called themselves the "trench coat mafia" hunted down fellow students and shot them.

I don't understand that mentality.

These tactics would have been used by Adolf Hitler in the death camps, or by one of the *Einsatzgruppen*, the Nazi death squads that cheerfully shot innocent civilians.

Human nature does not change. In all of our hearts lurk ancient temptations and eons-old evils. I guarantee that many people—too many people—would kill their neighbors if their government told them to do it. That is the danger of this type of thinking. Many people, if given a uniform and the authority, would round up innocent citizens who happen to belong to an unpopular political party or religious sect and kill them.

Our history is stained with examples of this.

Ask the Jews or the American Indians.

Don't assume that it couldn't happen again. It could. And I am telling you, when it appears, no matter what the symbol is, no matter the associated religion, you must understand that evil exists, and stand up to it and say, "No, that's not me."

It doesn't mean you have to surrender your religious beliefs. Evil men disguise themselves in the saintly robes of religion. It doesn't mean your religion is evil; it means he's evil. And so in casting out his evil, you make sure you yourself don't fall into that well.

Is Norway's suspected murderer Anders Breivik a Christian terrorist?

Some would say yes, and so would I. Not to slam Christianity, but as a warning to those of you who think that it can't happen within your own ranks. It can and it did, and it will happen again. Remember the words of Senator Barry Goldwater, speaking at the 1964 Republican National Convention, in San Francisco's Cow Palace: "I would remind you that extremism in the defense of liberty is no vice. Moderation in the pursuit of justice is no virtue."

I'm a Goldwater conservative. Those are the truest words ever spoken in my lifetime, next to Eisenhower's "beware the government military-industrial complex." However, it doesn't mean you have to slaughter innocent people if you believe that extremism in the defense of liberty is no vice. It doesn't mean you blow up parliament buildings. It doesn't mean you go out and kill children.

What I want you to understand is that you can be an extremist without being a murderer or terrorist. You can be an extremist in what you do, and work around the clock to save this nation without having to resort to what this lunatic did in Norway.

That's the whole point of my show, which is not to withdraw from the field of battle with multiculturalism, not to shrink from combat with socialism, not to retreat from fighting cultural and economic Marxism, but to redouble your efforts to fight them with everything in your power—but without resorting to violence.

Do not resort to terrorism, do not become a murderer, but be an extremist.

I'm trying to tell you, you can be an extremist without being a terrorist.

Being rational and being right doesn't often engender love from the people, but I have a job to do. My job is like Casey Stengel's: I've got to call them the way I see them, and I'm telling you this guy is a Christian and is a terrorist.

Accept the truth when you hear it.

But he's not the only terrorist among the religious, is he?

There are many others in many religions and I have pointed them out to you. We don't have to go through the Muslims, do we? They are too many to count. You should understand that there are Buddhist terrorists, Hindu terrorists, and in every country there are terrorists who wrap themselves in every religion.

So put aside the big lie that there was no Christianity in what he did. Well, maybe there's no Christian theology in what he did, I'll agree with that. Christian theology certainly doesn't teach that thou shall kill thy neighbor if you disagree with their politics. But in the age in which we're living, people try to find justification for their insanity and violence. The devil can quote scripture and too often he does. Certainly, he did in Norway.

Breivik was fascinated with the crusades and considered himself a member of the Knights Templar, a Crusader army of a thousand years ago. You can see this. Much can be said about the other side of the aisle, Muslim extremists and murderers like bin Laden and his number two, Ayman al-Zawahiri. Bin Laden was a businessman and engineer; Zawahiri was a medical doctor in Egypt. Neither was a theologian or member of the clergy, yet they self-identified as Muslims. It doesn't mean that all Muslims espouse what they espouse. Unfortunately, too many Muslims do, like ones who cheered after 9/11. Many of you don't want to accept that Breivik was a Christian or that any Muslims who say they don't believe in terrorism aren't telling the truth.

Many of you can rightly quote the Koran and its instruction that you shall smite an infidel and put a knife to his neck. It doesn't mean that all Muslims follow that radical or Salafi brand of Islam, although many of them do.

If you look at the Old Testament of the Bible, it has similar edicts. Most Christians don't read or follow the Old Testament (which includes the kosher dietary laws), but instead follow what they call the New Testament of Jesus Christ. But you won't argue that the poor old Orthodox Jew is walking around wanting to kill people because he believes the Old Testament. They understand that though it was God's word for those times, that doesn't mean it's God's word for these times.

I've heard Orthodox rabbis say to me: Who is it for us to decide which commandments we should or shouldn't follow? Where are you going with that one? Are you going to stone an adulterer? If you see your father's nakedness, you're supposed to slay the person who uncovers his nakedness?

Read Leviticus, which is unbelievably full of murder for moral offenses. I don't believe Orthodox Jews murder adulterers or homosexuals, do you? Therefore no one is following the Old Testament in every aspect or to the fullest degree.

How do you interpret this if you're a theologian? That's why I'm not religious. Do I believe in God? Sometimes I do, sometimes I don't. Sometimes I believe for days or weeks or years at a time, and some days I don't believe anything except in survival. The one credo I've found most operational and inspirational for me is survival. If that's religion, then I'm a religious man. That means survive, survive, survive and don't let anyone stand in your way, short of killing him.

On the other hand, I don't know if a religion is of any value unless it teaches you how to survive. I've felt that for a long time, going back to my belief system. I've been in temples, churches, and synagogues and I've never really felt at home in any of those houses of worship. It doesn't mean they're bad or don't serve a social purpose. They do. But do they teach people survival skills? Maybe they did for you, but not for me.

When I see stories of ancient Buddhist temples that taught young acolytes martial arts, so the monks with shaved heads living in the mountains of China and Tibet could defend themselves against invading armies using hoes and sticks, that is a religious teaching I can relate to. Is it a religious teaching to teach someone to fight with a hoe or a fishing rod or a chair? They decided that the young men shouldn't go to the gas chambers like the Jews did in World War II. I respect that of the Buddhists who teach their men to kill the enemy. If the Jews did that at the beginning, there wouldn't have been a Holocaust. I've studied it inside and out. I don't care how many times you quote the Bible, what good is it if a man comes to kill you, your wife, and children?

The Jews who went into the sewers and woods of Poland are the Jews I respect. They cut Nazi throats, bombed their rail lines, and burned their supply depots. Those are the Jews I respected, not the Jews who sat begging for their lives and got only slaughter and humiliation in return.

With Breivik we're talking about the slaughter of innocent children, albeit the children of the Labor Party, at the core of Norway's cultural identity. But they were unarmed teenagers. Where's the nobility in Breivik's act? How can anyone say it was justifiable? How can you understand this? You don't go kill children because they belong to the Labor Party.

Those of you who served in Iraq know this. They tried to kill as many children as possible. Why would Muslims want to kill Muslims? I understand why they'd kill American soldiers—I hate them for it and would kill them for it—but what's their thinking? Their twisted logic is that unless you're an extremist, you're a collaborator—then, you side with the crusaders, the Americans.

That's the same twisted reasoning the psychopath Breivik employed in Norway. The left is trying to say he was a student of Timothy McVeigh and the Unabomber, but he was a student of al Qaeda. He copied their tactics: He attacked the innocent members of his race and religion, hoping to polarize society and spark a civil war. Al Qaeda, as shown in its manuals captured in Afghanistan, has the same methods, dreams, goals, and tactics.

The secret society that threatens every one of us extends far beyond even the vast star-packed universe of the Government-Media Complex. In the next chapter, I will show you how Soros, Obama, and the rest of a secretive financial elite control the United States economy and are purposefully driving it to destruction. In the end, you could lose your job, your house, your car, and even your country. Their subversion of the media, which acts as sword and shield to protect them for public scrutiny, allows them to carry out a much larger, far-ranging plan.

CHAPTER 4

Tyranny of the Treasurer

Obama and the Leninist Economic Vision

Vladimir Ilyich Lenin had a disturbed vision. He thought that a nation could only grow more prosperous when it was controlled by a vanguard, an elite. The Leninist elite would lead the Russian people to a world free of pain and poverty. No more haves and have-nots. No more private property. Only boundless prosperity.

All Russia had to do was transfer its entire wealth to Lenin's elite.

The Leninist vision had terrible consequences. If you didn't want to relinquish your property, Leninists would take it. And instead of boundless prosperity, the Russian people endured unthinkable poverty and famine. Photographs abound of groups of men, women, and children, their skeletal bodies marked by grotesque swollen bellies, sitting helplessly at the edge of a road or in front of the empty huts they once called home, waiting to die. Cannibalism was rampant, so much so that the Russian government was forced to issue edicts against it.[1]

If you don't believe it could happen here, listen to the president. At a million-dollar fund-raiser in San Francisco in October, 2011, Obama warned his audience that if he's not elected in 2012, Americans will have to become more self-reliant than they are now. His exact words:

The one thing that we absolutely know for sure is that if we don't work even harder than we did in 2008, then we're going to have a government that tells the American people, "you are on your own. . . . If you get sick, you're on your own. If you can't afford college, you're on your own. If you don't like that some corporation is polluting your air or the air that your child breathes, then you're on your own." That's not the America I believe in. It's not the America you believe in.[2]

The America that Barack Obama "believes in" is the same America Vladimir Lenin would have believed in.

I've warned you repeatedly that in his first three years in office, Barack Obama has done everything he can to bring the Leninist vision to America. He's engineering the takeover of the world's largest economy by his own central government, which is determined to convert our country from the most powerful and successful democracy on the planet into a third-rate nation.

Who controls the United States under Obama's Leninist vision of what this country will become?

I call it the financial aristocracy.

They're the modern equivalent of Lenin's elite.

I'm the only one who's spoken out against the new financial aristocracy.

That's because it's made up of Wall Street power banks like Goldman Sachs, JPMorgan Chase, and Bank of America, and most talk show hosts and commentators and news organizations—and especially our politicians—are afraid to take them on.

The group also includes anti-American financiers like George Soros, the International Monetary Fund (IMF) and the World Bank, Obama's egghead college-professor financial advisors, and government agencies of the United States and Western European socialist nations.

Like the Leninist elite, they've conned the world into believing that they're looking after the interests of "the people" when in fact they're in the process of seizing control of the world's financial assets at the people's expense.

Thanks to the financial aristocracy, you're no longer working so you can afford the things you need to live a comfortable life and pass along an inheritance to your children; you're working so you can give your money to these financial oligarchs to underwrite the enormous debt they're incurring in order to advance their takeover of the global economy.

What's the most significant result of Obama's attempts to reduce our economy to second-class status?

Our unprecedented failure to recover from the 2008 recession.

Look at the numbers. The Obama recovery has produced average GDP growth of only 2.4 percent over the two years since the recession ended. That's the most pathetic recovery in U.S. history. For the third quarter of 2011, the U.S. economy grew at a 2.5 percent annual rate—and the administration celebrated.

Let me translate that into real terms for you: When compared to the 6.2 percent *average* GDP growth following the other serious recessions since World War II, the Obama recovery has been so weak that it's cost us nearly a trillion dollars in lost GDP.[3]

The Obama economy has also cost us as many as *10.5 million* jobs. In 2011, as the recession dragged on, more than 14 million people were unemployed. For every job opening, there were seven people who needed a job.[4]

It's what happens when you put your economy in the hands of a socialist like Obama, who is committed to proving that capitalism doesn't work.

Did you know that not only is Obama determined to bankrupt the United States, he's out to destroy the middle class in the bargain? I've told you over and over that when Obama attacks Wall Street he's only trying to divert attention from his own partners in crime: the very banks and big businesses that he is in league with as he pushes his economic takeover.

His real target is the middle class.

Don't get me wrong. I'm no friend of big banks. Banks aren't in the business of standing up for the people. They exist to make a profit like other corporations.

But they're not supposed to side with Leninists to steal our wealth.

Like Lenin, Obama is determined to confiscate wealth and redistribute it. He's not going to redistribute it to the people, though. He's going to redistribute it to the financial elite who are rapidly gaining control of our economy.

Obama's henchmen are doling out some of the money they seize to the increasing number of poor people in the United States, but only enough so they can stay poor. By the time they're done running up debt, printing money, and raising taxes, all but the wealthiest will be reduced to poverty and dependence on our government. In the process, the middle class—better known to the Leninist sympathizers who make up Obama's financial team as the bourgeoisie—will disappear, as it has done in every big-government socialist economy in the world.

It's no secret that Obama, like all leftists, despises the bourgeoisie and all it stands for.

I describe the economic philosophy of the Obama administration as "sustainable bankruptcy."

Do you realize what the intellectual idiots who control our finances have caused?

Yearly trillion-dollar-plus deficits; a Social Security fund that has run out of cash; millions of homeowners either underwater with their mortgages or already in foreclosure; a capitalist crony system that is churning out billions in profits simply by gaming the system through creating nonexistent collateralized debt; inflation on the rise; the IRS empowered to expand their crackdown on taxpayers into the health insurance field; our own government investing billions in foreign energy projects while America's energy industry is hamstrung by our own leftist government; more Americans than ever in history reliant on the government for their income.

We're living Obama's new American economic nightmare.

When you hear what Obama's Agriculture Secretary Tom Vilsack has to say about the increase in people dependent on the government, you'll start to understand why: "[W]hen you talk about the SNAP program or the food stamp program, you have to recognize that it's also an economic stimulus. If people are able to buy a little more in the grocery store, then someone has to stock it, shelve it, process it, package it, ship it. All of those are jobs. It's the most direct stimulus you can get in the economy during these tough times."[5]

That's the Obama idea of how the economy works: Redistribute income. Give people something for nothing. Call it job creation.

As this graph shows, under Obama the percentage of people living in a

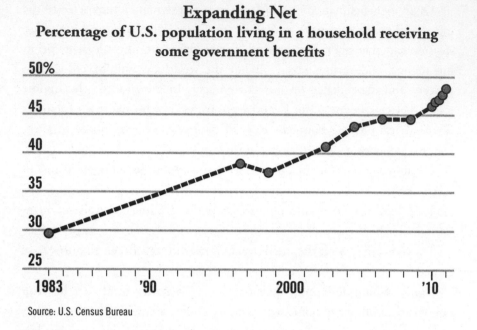

Expanding Net
Percentage of U.S. population living in a household receiving some government benefits

Source: U.S. Census Bureau

household receiving government benefits is projected to rise nearly ten percent before Obama leaves office in 2012, after having risen only about ten percent over the previous 25 years:

Nearly half of all the people in the United States live in households that are receiving direct government benefit payments. That's up by nearly ten percent since Barack Obama assumed power and began to impose his economic vision on America.[6]

It's already happening. But not quite the way you think.

Communist economies never transfer wealth to the poor and middle classes. They transfer it to government officials and their cronies. That's what's going on right here in the United States.

In the top-earning 1 percent of American households, income grew by 275 percent over the last three decades. Middle-class incomes? Less than 40 percent.[7]

It's just what the president ordered.

Obama's intent is not just to remake America in the image of the failed socialist economies of the twentieth century. His ultimate goal is much broader: to weaken the U.S. economy to the point where it is just another player in a new world order controlled by a political/financial elite.

Obama has a grievance against this country, and he's determined to bring America to its knees, to make it pay for its "imperialist" past. I see the Obama administration's economic tactics—the huge increases in the federal debt, the unchecked spending, the weakening of the dollar, the class warfare—as part of a larger pattern. They show me that despite the fact that he criticizes Wall Street and big business every chance he gets, Obama is in league with the financial and corporate elite to do nothing less than cede American power to them as part of his overall aim to weaken our country.

The Obama Certainties: Debt and Taxes

In order to advance his Leninist vision, Obama added $4 trillion to our national debt during the first three years of his presidency, more than George W. Bush added in his eight years in office.[8]

You can't say I didn't warn you.

In January 2009, before he was inaugurated, Barack Obama announced his spending plans, then rammed the largest domestic spending initiative in history through Congress. In the spirit of Lenin, he promised it would *create* three million American jobs. Instead, he transferred money that should have stayed in the private sector into "shovel-ready" infrastructure projects, protecting the jobs of teachers and other unionized state and federal employees, promoting "green energy" jobs, and rewarding states for their irresponsible fiscal policies by giving them money to keep operating without cutting back.

Like Lenin's promises, Obama's weren't kept.

Since Obama took office and spent upwards of a trillion dollars in stimulus money, we've *lost*, depending on who you talk to, anywhere from three million to ten million jobs.[9]

Do you seriously think that Barack Obama intended to create three million new jobs? Or that he wants to see the unemployment level go below nine percent, where it's been for all but two months of his presidency? Do you think he wants to see Americans back at work?

The stimulus plan that was supposed to put Americans back to work was a phony from the start.

Are you aware that the stimulus plan was written by Marxists?

America-hater and Obama-puppetmaster George Soros's Apollo Alliance conceived and designed the stimulus plan. It was never designed to jump-start the U.S. economy or to create jobs. It was designed to transfer wealth and power from the people of the United States to the increasingly powerful central government. Despite the trillion dollars in spending it authorized, unemployment has remained officially at nine percent, but practically it stands somewhere between 16 and 20 percent.[10]

In fact, Obama was actually trumpeting that fact.

On the Sunday before Obama's September 8, 2011, jobs speech, the Associated Press, part of the Government-Media Complex, released a story saying that the real unemployment rate was not 9.1 percent, as was widely reported, but 16.2 percent when you factored in people who were working part-time but wanted full-time employment and those who had become discouraged and dropped out of the labor market altogether.

Why would one of Obama's publicity outlets release such information?

I'll tell you why.

The 16.2 percent figure is exactly what Obama wanted.

It is not now and never has been Obama's intent to reduce unemployment. Unemployment benefits are one of the ways he keeps people on the government dole. The longer they're extended, the less incentive people have to look for work and the more control government has over their lives.

The release of this information was part of the run-up to the president's jobs address. Unemployment numbers that high justified Obama calling for another $447 billion in stimulus spending.

Two weeks after the end of his Martha's Vineyard stay, Obama finally explained that his economic team had come up with a "new" plan. The plan consisted of pork barrel projects designed to subsidize labor and public employees unions by adding infrastructure projects and upgrading 35,000 school science labs so educators can teach our kids how to put a condom on a cucumber more effectively. Obama also called for extending unemployment benefits to make sure the unemployed don't have to get up off their couches and actually look for work.[11]

That new plan was never intended to be enacted into law, or even voted on by Congress.

It was Obama's way of trying to make Americans believe that Republicans were opposed to job growth. It was nothing more than a phony campaign promise meant to make the other guy look bad.

America, reach for the lubricant. The most corrupt president in American history is not going to let up in his campaign to bankrupt this country, and you'll be the next victims.

The Debt Ceiling Junta

Let me tell you how Obama has managed to fund his economic takeover. He hasn't done it legally, through submitting and passing budgets. In fact, Obama's 2012 budget would have added another $1.5 trillion to our debt. It was rejected *unanimously* by the Senate!

No, he's not stealing us blind through submitting budgets, he's doing it through continuing spending resolutions. The closer we got to the debt ceiling, the closer we got to a new opportunity for Obama to expand his economic takeover—by the establishment of a junta.

By mid-2011, thanks to the largest spending binge in our nation's history, we had reached the $14.2 trillion U.S. debt ceiling. Congress finally passed a bill to raise the debt ceiling after Wall Street co-conspirators, led by Goldman Sachs CEO Lloyd Blankfein, released a letter warning that the debt ceiling had to be raised or we faced catastrophic consequences. These are the same financial hypocrites who contributed heavily to getting Obama elected and are now profiting through financing the welfare state Obama's policies have done so much to expand.[12]

The deal that resulted from the debt ceiling crisis raised the $14.2 trillion federal debt limit by a total of $2.4 trillion to about $16.6 trillion. Raising the debt ceiling would take place in two stages.

Half of the $900 billion first stage took effect on August 2, when the deal was signed. The first stage of the debt ceiling increase was granted in exchange for $917 billion in spending cuts over the next decade. The initial amount raised the debt limit to $14.694 billion.[13]

On August 3, the day after the legislation was signed, the U.S. debt shot up by $239 billion, to $14.532 trillion, putting us within $160 billion of the new temporary debt ceiling.[14] By mid-November 2011, we were more than $15 trillion in debt.[15]

It's the second stage of the deal that I see as contemporary Leninism at its most blatant. It requires a "select" 12-member committee of six Democratic and six Republican legislators from the House and the Senate to find another $1.2 trillion in spending cuts before the end of 2011.

This amounts to nothing less than the equivalent of a Russian junta. The only difference is that this committee was chosen to do the opposite of what Obama said it was for. It was chosen in order to fail to come up with a plan.

There's no precedent, no constitutional provision for this illegitimate body. Even leftist Democratic representative Maxine Waters said that the supercommittee "threatens our democratic process with its unconstitutional structure."[16]

It also threatens to undermine the U.S. military, another of Obama's aims in his takeover. If the junta fails to agree on budget cuts, Obama's aim to weaken the U.S. military will be advanced significantly, because $600 billion of the additional $1.2 trillion in savings will come in the form of military spending cuts that kick in automatically. California representative Buck McKeon said, "If enacted, the trigger wouldn't just gut our military. It would put it out of business. . . . [I]t could mean the end of the United States as a superpower, resigning us to a multi-polar world and all the dangers that come with it."[17]

The remaining $600 billion in cuts will come from nondefense spending, including cuts to Medicare providers.

In fact, though, it turns out that this committee's deadline of November 23, 2011, was another phony one. The drastic cuts to the military aren't scheduled to kick in until 2013, well after the 2012 elections. Even so, the lame-duck Congress will rescind those automatic cuts to the military before the end of 2012. Even a committed anti-American leftist like Defense Secretary Leon Panetta wouldn't let cuts to our national defense like the ones Obama would like to see go into effect, much as the president would like them to happen.

Obama's decimation of the military goes beyond his cutting its budgets, though. Do you remember the incident in which four American Marines allegedly urinated on terrorists they had killed? The military's response speaks volumes about our attitude toward the defenders of our nation, and I addressed the issue in a *USA Today* OpEd piece. Here's what I wrote:

Most of the "men" in the media—the same journalists who are orchestrating the attack on American Marines who allegedly urinated on Taliban terrorists they killed in battle—have never even been in a fistfight, let alone a firefight.

And yet here they are, piling on the real men who have the guts to go into combat against 15th-century radical Islamist throwbacks, to face a barbaric enemy eye to eye, blade to blade, bullet to bullet.

The piling on our Marines by the media has done far more damage to the morale of our fighting men than our soldiers' actions.

And yet we hear Marine Corps Commandant General John Amos calling the act "inconsistent with [our] high standards of conduct," then whining about our commitment to "upholding the Geneva Convention, the laws of war and our own core values."

This in the face of Taliban atrocities—car-bombings, beheadings, the murder of women and children—that flout the Geneva Convention and yet are routinely ignored by the media and our military higher-ups, who also downplayed the Ft. Hood massacre by Major Nidal Hasan.

Our fighting men must be supported at all costs at a time like this. How can we in good faith ask them to put their lives on the line against the vilest, most extreme of enemies while at the same time we attack them in the press at every turn?

I'm not saying that what these Marines did isn't a violation of military code. But unlike murderers who are granted appeal after appeal and for whom liberals in the legal system bend over backwards to avoid punishing, these Marines have already been tried and found guilty in the media.

The desecration of war dead is universally condemned, but it is as old as war itself. These boys are not the devils. The Taliban are the devils.

Our Marines are heroes. They should be rewarded for having the guts to go into combat, not punished for stepping over the line.

It's the latest attempt by the liberal media and the increasingly timid military brass to destroy the few, the proud, the brave, the Marines. "To err is human, to forgive divine."

Taxation Without Representation

Obama's Leninist tactics continue with his seizure of American wealth, though his method of seizing property is less direct than Lenin's.

Lenin simply sent his thugs out to seize the property of Russian citizens who didn't want to give everything to the ruling party. Agents of the Soviet government would enter a village and systematically take all the meager possessions of its inhabitants. These included virtually all clothing, from jackets and trousers and dresses to shoes and winter underwear. Household items, from kitchen utensils to bedding, were also seized, leaving the peasantry with, literally, not so much as a pot to cook in.

Obama doesn't send his thugs directly into Americans' homes; he simply taxes us into poverty.

Since he took office, Obama has added 21 *new taxes* on Americans.

Many of them are part of the Obamacare legislation.

Did you know that as of 2014, if you're not buying "qualifying" health insurance, a family with two adults will pay $190 or one percent of its adjusted gross income (AGI) in increased taxes?

By 2016, that number will increase to $1,390, or 2.5 percent of AGI!

That's just the beginning of the president's wealth confiscation plan.

If you're a small business owner with 50 or more employees who does not offer health insurance through your company and at least one of your employees qualifies for a health tax credit under Obamacare, you'll have to pay an additional nondeductible tax of $2,000 for every one of your employees beginning in 2014. If any of your employees receives health insurance through the Obamacare insurance exchange, your tax rises to $3,000 per employee.

If you're one of the "wealthy" Americans making $200,000 annually—$250,000 for a family—you'll have to pay an additional tax on your investment income starting in 2013. This means that your top tax rate on investment income would rise from 15 percent on long-term capital gains to 23.8 percent. For dividends and other investment income, the top rate would rise from 15 percent to 43.4 percent.

Obamacare also includes tax hikes for the following: Medicare payroll tax, Health Savings Account withdrawals, companies that manufacture medical devices, people who suffer from serious medical conditions and itemize their medical expenses on their tax returns, people with special needs children, retirees who get drug benefits as part of their pensions, charitable hospitals, companies that develop new drugs, and health insurers, among many others.[18]

I call that confiscation.

Do you understand that it's nothing more than another part of Obama's Leninist takeover strategy?

In Obama's America, as it was in Lenin's Russia, certain people are given favored treatment based on their standing and influence with the government. We're all familiar with the Obamacare exemptions, which grant waivers from complying with the law to more than 1,000 of these favored corporations and labor unions. Nearly 20 percent of the 204 waivers approved in April 2011 were granted to "gourmet restaurants, nightclubs, and fancy hotels" in California congresswoman Nancy Pelosi's district.[19] Fully half of all the people covered by the waivers are members of labor unions.[20]

The administration is simply excluding these political favorites from having their money confiscated through Obamacare.

The political favoritism doesn't end with Obamacare waivers.

Obama's capitalist cronies are also exempted from having to pay corporate income taxes in many cases. General Electric's Jeffrey Immelt is the prime example.

In a blatant payback to GE for its political support of the president's policies, Obama appointed Immelt to the position of Jobs Czar—head of the "President's Council on Jobs and Competitiveness." Immelt defined his mission as making the commission "a sounding board for ideas and a catalyst for action on jobs and competitiveness." He's done that by making our economic adversaries more competitive through outsourcing American jobs overseas and continuing to undermine job creation in the U.S.[21]

Under Obama's and Immelt's leadership, the United States has dropped to fifth place in economic competitiveness among the nations of the world, according to the World Economic Forum. In 2008, we ranked first.[22]

Immelt is no friend of the United States. In fact, Immelt's business dealings favor China over the U.S., and the president's appointment of Immelt may actually be another way Obama helps China gain a competitive advantage in its economic war with the United States. It's a reward for political crony General Electric, whose political lobbying arm, GEPAC, donated $2.4 million to Democrats in order to get cap-and-trade legislation passed. As one GE executive's e-mail explained, "If this bill is enacted into law it would benefit many GE businesses." With the bill set to impose new greenhouse gas emissions standards for aircraft engines, and with GE being one of the primary companies in America producing the engines, the benefits are obvious. In addition, GE received $75 million in taxpayer money to build "smart meters" that remotely monitor electric power use.[23]

Once Obama's cap-and-trade bill failed to pass, there was reduced demand for Immelt's jet engines here, so the new Jobs Czar struck a deal with the Chi-Coms to manufacture engines in China, putting American aviation know-how into the hands of our biggest creditor and one of our strongest economic and military foes. This treasonous technology transfer followed Immelt's announcement that he was moving GE's 115-year-old X-ray business headquarters from Waukesha, Wisconsin, to Beijing as part of GE's "Spring Wind" initiative to develop more medical products and services in China.[24] GE's investment of $2 billion in China will include hiring 65 engineers and additional support staff in its new China facility.[25]

Microsoft mogul Bill Gates seems to be taking a lesson from Immelt. Gates is in discussions with the China Nuclear Corporation to have his company, TerraPower, locate a manufacturing facility in China and co-operatively develop Generation IV nuclear power reactors there.[26]

Under jobs czar Immelt's leadership, GE has laid off 21,000 of its American workers and closed 20 of its factories from 2007 to 2009. Less than half of GE's workforce remains in the U.S.[27]

The stenographers who pose as journalists in the new Obama Government-Media Complex never bothered to question either Immelt's appointment as jobs czar or the transfer to China of the very jobs he is supposed to protect.

They haven't questioned GE's tax status, either. Although the company declared $14.2 billion in profits in 2010—up 77 percent from the previous year[28]—more than $9 billion were earned offshore and were not subject to the 35 percent U.S. corporate tax rate, the highest in the world.[29]

GE wasn't taxed at the 35 percent rate.

GE wasn't taxed at all for its U.S. earnings.

In fact, not only did GE pay no taxes on the $5.1 billion it earned in the United States in 2010, but it actually got a *$3.2 billion corporate tax credit.*

GE maintains the largest and best corporate tax department in the world, and manages to navigate the maze of corporate tax regulations better than any other multinational company. It also manages to buy the support of the Obama administration as one of the president's favored crony capitalists.

Did you think the corporate giant got away with murder in 2009? Its 2010 corporate tax return was 57,000 pages long!

Its profit was $14 billion, nearly three times the year before.

And once again GE paid no corporate taxes.[30]

Immelt's GE benefits immeasurably from Obama's selective Leninist tax strategy, which is built on providing political favor to the administration's allies. That Obama would even consider appointing an anti-American, anti-jobs mogul like Immelt to the position of jobs czar should be an impeachable offense.

Immelt and General Electric aren't alone in their ability to evade taxes. Dozens of other companies, from Pacific Gas and Electric to Wells Fargo, paid no federal taxes for the past three years. Wells Fargo also received a number of tax subsidies. Hewlett-Packard, Yahoo, and Levi-Strauss paid taxes at less than half the 35 percent corporate tax rate. In all, the companies listed for tax avoidance in a report issued by the Citizens for Tax Justice received more than $200 billion in taxpayer-funded subsidies.[31]

The crony capitalism doesn't stop with Immelt's position as Jobs Czar. Lewis Hay, the chairman and CEO of green energy company NextEra Energy, sits on the president's Council on Jobs and Competitiveness. His company has received nearly $2 billion in federally guaranteed loans. The situation is no different from Immelt's. It's a clear conflict of interest.[32]

But the Obama administration's crony capitalism goes way beyond simple conflict of interest and avoiding corporate taxes.

Do you understand that it's now bordering on treason?

Obama's science "czar" is a committed leftist named John Holdren. He's one of about 40 Obama czars—nobody knows the exact number—more than Lenin's government ever had. Holdren's the traitor who said that the Intergovernmental Panel on Climate Change (IPCC) produces the "most important conclusions" about climate change.[33] He's also committed treason by giving U.S. technology to China. Holdren has visited China several times in order to participate in talks about having NASA cooperate with the Chinese People's Liberation Army, which controls China's space program. In doing this, Holdren had defied legislation that specifically prohibits this kind of information exchange.[34]

Do you understand what I'm saying?

Holdren is committing what amounts to treason under U.S. law.

Giving U.S. technology to China—as both Holdren and Immelt are doing—is part of the Obama administration's attempts to undermine the private economy, and it's unprecedented in its scope.

It's driving us rapidly toward bankruptcy.

The list of economic failures includes the first debt downgrade in U.S. history, and the highest budget deficits, federal spending, and federal debt as a percentage of GDP since World War II. Under Obama home ownership has declined to its lowest level since the 1960s. The number of Americans paying taxes is the lowest in the modern era, while those dependent on the government for aid is the highest in history.[35]

On top of this, Obama has announced that he will veto any bill containing spending cuts that are not accompanied by a $1.5 trillion tax increase on the "wealthy." This is the so-called Buffett Tax, named after multibillionaire Warren Buffett, who traded his support of Barack Obama's millionaire tax and Obama's assurance that the Fed would not allow BofA to fail for a chance to buy Bank of America preferred stock that will net him billions over the next few years, even as the bank flounders.

Warren Buffett has repeatedly claimed that he pays less in taxes than his secretary and that millionaires like him should pay their fair share of taxes. As I've told you, the reason Buffett pays a small percentage of his income in taxes is that he takes a very small salary—less than $100,000 annually by some reports—as income. The rest of the proceeds from his investments stays in his company, where his tax accountants take advantage of every loophole to keep their tax bite very small.

Like every one of Obama's policy initiatives, Obama's millionaire tax is based on false information. In this case, Obama is insisting that "millionaires pay a lower percentage of their incomes in taxes than their secretaries do." They're taxed, Obama says, at a much lower rate than the middle class.

In general, ordinary millionaires pay much more than the rest of us in taxes as a percentage of their income. People whose incomes are higher than $1 million annually pay about 29.1 percent in taxes, while people with incomes between $50,000 and $100,000 pay about 15 percent.[36]

Warren Buffett's income from wages falls in the second category.

People who earn most of their income from their job tend to pay taxes at a higher rate than those who earn a high percentage of their income from investments. But the differences still don't result in millionaires paying a smaller percentage of their income in taxes.[37]

It's just more of the same lies Obama repeatedly tells in order to perpetuate class warfare and his assault on "the rich."

Most Americans think that America's economic failures are occurring because the Obama economic team is blindly committed to Keynesian economic policies and that the president's economic team doesn't realize—or won't admit—that Keynesian economics simply doesn't work.

I don't buy that.

I think the situation is much more serious.

I have no doubt that the Obama administration's intent is to weaken our economy in order to further a crisis of the international financial system and facilitate its takeover by a U.S./global financial elite on the scale of the Leninist postrevolutionary takeover of the Russian economy.

I'll explain what I mean.

Roots of the Current Crisis

In *Trickle Up Poverty*, I showed you how the stock market crash of 2008 was engineered in large part by hedge fund managers in order to insure that Barack Obama was elected president. Just before the economic collapse, the largest Wall Street banks came running to Washington, begging to be spared the consequences of the risky investments they'd made. Treasury Secretary Hank Paulson handed them hundreds of billions in Troubled Assets Relief Program (TARP) money to cover losses they were incurring because of their investments in risky housing derivatives and collateralized debt obligations.

What wasn't revealed at the time was that that money didn't scratch the surface.

Did you know that after TARP, the Federal Reserve secretly loaned another $1.2 trillion to U.S. and international banks in order to avoid a further catastrophic meltdown of the global financial system?

Take a look at some of the financial institutions involved:

Morgan Stanley: $107 billion
Citigroup: $99.5 billion
Bank of America: $91.4 billion
Royal Bank of Scotland: $84.5 billion
Goldman Sachs: $69 billion
JPMorgan Chase: $68.6 billion[38]

As one official put it, "These are all whopping numbers."[39]

It took three years for the numbers to come to light, but during that time—despite the fact that they would have gone bankrupt without the Fed loans—I didn't see the banks in question slowing down their suicidal practices.

Instead, they continued to engage in high-risk lending and hedge-fund trading. Even though our country is on the brink of bankruptcy, they're still following the same strategy and pushing the same agenda today. Their ongoing treachery continues to put the savings and livelihoods of American middle-class citizens at risk, because it's the middle class they're counting on to bail them out again.

The treachery is going to be hard to prove, though.

The Securities and Exchange Commission (SEC) has been accused by one of its employees of destroying more than 9,000 files related to investigations of the very financial giants—Goldman Sachs, Lehman Brothers, Citigroup, Morgan Stanley, Wells Fargo, and Bank of America—who played a large part in causing the 2008 crisis and who benefited enormously from the TARP bailout and loans from the Fed.[40]

I'm the only one I know of who's telling you that the secret loans made to these financial giants as part of the Obama administration's strategy are for the purpose of transforming the U.S. capitalist economy into a managed economy, controlled by the government in league with the world's largest financial institutions.

As big as the numbers I'm citing are, the "official" U.S. federal debt represents only a fraction of our total debt obligation. Unfunded liabilities—what we've promised to pay to retirees in the form of Social Security and Medicare benefits—add an additional $100 trillion to our debt. The Social Security trust fund had accumulated $3 trillion from those paying into it, but the federal government has borrowed every penny of that and issued "special-interest bonds" to Social Security, so there's no cash available in the trust fund.

Add in another several trillion dollars for other unfunded liabilities such as federal employee and military retirement benefits and federal guarantees through such agencies as TARP, Fannie Mae, and Freddie Mac, along with the projected Obama 2012 budget deficit of more than $1.5 trillion, and you start to get an idea of our true economic condition.[41]

Our national debt and unfunded liabilities amount to nearly ten times the U.S. annual GDP, and there's no sign the know-nothings in Congress have the will to do anything about it.

But let me give you the larger picture of what's happening.

It goes way beyond the thievery committed by the Obama administration and the crony capitalists who run America's banks and brokerages.

The annual world gross domestic product—in other words, when you add up the GDPs of every nation on the planet—amounts to about $60 trillion. The United States, even though we have less than six percent of the world's population, generates 25 percent of the world GDP, or about $15 trillion.

Keep those numbers in mind when you read what I'm about to say next.

I've explained to you in *Trickle Up Poverty* how financial crooks like George Soros engineered the economic crash that carried Barack Obama to the presidency. One of the reasons that crash happened was financial "derivatives," debt instruments created in order to spread the risk of loss, especially in the mortgage market. The reason we haven't been able to recover from the current recession—and the reason that European banks and sovereign governments haven't been able to stabilize the risk of economic collapse in Greece, Italy, Spain, Portugal, and Ireland—is that almost every big bank and every government carries those derivatives on its books.

Do you know what those derivatives are worth?

Six hundred trillion dollars.

Ten times the GDP of the entire planet.

More money than the world generates in ten years by legitimate economic activities like manufacturing, services, and sales.

The world economy is a prisoner of the debt instruments the big banks and sovereign governments created, and there's no way to avoid it.

There's not enough money in the world to stave off the coming economic crisis.[42]

Who's on the hook for all this?

They've rigged the game in their favor to the point where *U.S. taxpayers are responsible for the crooked dealings of the banksters who have taken over the U.S. and global economy and are using it as their ATM.* They're getting rich beyond imagination by manipulating every single market in the world—from commodities like gold, silver, oil, and corn to the bonds that sovereign nations use to finance their debt.

The large banks and brokerages like Goldman Sachs call the financial shots around the world, taking risks that banks weren't allowed to think about only a little more than a decade ago, and when they lose, American taxpayers are on the hook to bail them out, thanks to the crony capitalist in the White House.

The Goldman Sachs Revolution

Over the past several decades, I've watched banking giant Goldman Sachs develop the strategy that is now being played out as the international economy comes more and more under the control of a small group of giant financial institutions. The instigator was former GS CEO Robert Rubin.

When Rubin left Goldman Sachs in the mid-1990s to join the Clinton administration, becoming Treasury Secretary in 1995, he began a parade of GS executives through the cabinets of American presidents that's still happening. Virtually all the key players on the Obama economic team are former Goldman Sachs employees or sympathizers, and many of them, including Treasury Secretary Timothy Geithner, former senior economic adviser Lawrence Summers, and former budget director Peter Orszag, were mentored by Rubin himself.

The *Times* of London has named this unholy triumvirate the "Robert Rubin Memorial All Stars."[43]

After he left the Clinton administration, Rubin, as chairman of another banking giant, Citigroup, brought the Ponzi scheme that he had begun to set up during his tenure as Treasury secretary to the private sector. It's become the model for the one Goldman Sachs, along with the Bilderberg Group, the IMF, and the Council on Foreign Relations, is now implementing on a global scale. In 2008, a lawsuit was filed against Rubin for defrauding Citibank shareholders of some $122 billion, an amount greater than twice that of now-jailed Ponzi schemer Bernie Madoff.[44]

Another Goldman Sachs alum, Jon Corzine, pulled the same kind of scheme on his investors. Corzine, after he left as chairman of Goldman Sachs in 1999, was a U.S. senator from New Jersey before he became the state's governor in 2006. He fleeced New Jersey taxpayers and lined the pockets of his union cronies until Chris Christie unseated him in 2009. After he was ousted from the New Jersey governor's office, Corzine went back into the financial services business, heading up a company called MF Global Inc., a company specializing in futures trading.

Less than two years after Corzine took the helm, the company filed for bankruptcy after misappropriating customers' money and using it to make $6.3 billion in bets on European bonds. As a result of the turmoil in the Eurozone sovereign bonds—the same ones I've just told you will be the basis of the next crash—the company was downgraded by credit rating agencies to "junk" status. In the process, it was discovered by "regulators"—the same people who could have prevented the debacle if they'd kept a closer watch on the company and its CEO in the first place—that MF Global had violated a cardinal investment industry rule: They failed to "segregate" their clients' money from the company's own cash. Somewhere in the neighbor-hood of $1.2 billion in clients' money is missing and unaccounted for.[45]

More than 1,000 MF Global employees were summarily fired. Corzine resigned in the midst of the growing scandal, then hired Andrew Levander, an attorney who specializes in defending those accused of committing the same kinds of white-collar crimes Corzine is accused of.

Corzine was only one in a long line of crooks who shuttled back and forth between government and the financial industry using what they'd learned at Goldman Sachs.

But whereas Madoff and Corzine focused on fleecing individual inves-tors, Rubin developed the model for the institutional rip-off. What we're currently witnessing in the international financial markets is Rubin's model put into practice on a global scale.

Here's how it works.

The current "debt crisis" is a central part of the scheme. The U.S. Trea-sury continues to print money and sell Treasury bonds in order to finance government spending and disguise the extent of the potential damage from the near-worthless debt obligations the government and big banks still maintain on their balance sheets.

The people on the hook for this debt?

The American taxpayers.

The same people they've been fleecing since Obama took office.

I see it as the perfect setup: a captive group of investors at the mercy of a ruthless and desperate government financial power elite abetted by politi-cians who are unwilling or unable to see what's taking place before their very eyes.

The financial industry power players control the action.

The Goldman Sachs/U.S. government Ponzi scheme is carried on at a very high level. For starters, Treasury Secretary Geithner has the entire U.S. GDP and the "full faith and credit of the United States" at his dis-posal. What he and his Obama administration cohorts, along with their Wall Street cronies, are trying to engineer is in every way the same fraudu-lent scheme that Bernie Madoff and Robert Rubin pulled off against their investors.

In this case, though, it's American taxpayers who are the "investors" being scammed, and we don't have a choice in the matter. If we don't pay our taxes, we'll be prosecuted to the full extent of the law.

It's a con man's dream.

There are others involved in this conspiracy.

In *Trickle Up Poverty* I explained how the Bilderberg Group, a cadre of international power players, has been designing a strategy to take over the world economy, creating a two-tiered class system consisting of "elites" and everybody else. You and I are in the second group, and if the Bilderbergers have their way we will surrender control of our financial destiny and become beholden to them for everything we need as they move toward creating a world financial system which they dominate. The Council on Foreign Relations (CFR) is, along with the Bilderberg Group, a second key player in the attempt by financial elites to rule the world's economy.

In 1999, Bill Clinton—a member of both the CFR and the Bilderberg Group—along with his Treasury secretary, Robert Rubin—a member of both groups and a former Goldman Sachs CEO—pushed through the passage of the Gramm-Leach-Bliley bill, also known as the Financial Services Modernization Act of 1999. Gramm-Leach paved the way for the current economic crisis by repealing the Glass-Steagall Act.

Glass-Steagall had been passed in 1933 in order to spur our recovery from the Great Depression. The law required the separation of commercial banking operations from investment banking in America's financial institutions. Prior to Glass-Steagall, banks were permitted to loan money as well as to invest it and broker investments for their clients. That led to their being dangerously overleveraged and to one of the most disastrous effects of the stock market crash of 1929: the failure of American banks.

By 1933, four years after the Great Crash, nearly half of the more than 25,000 U.S. banks had failed. Americans' savings disappeared.[46]

Prior to Glass-Steagall, banks had been essentially unregulated, and they were allowed to engage in both commercial and investment banking activities. Glass-Steagall separated commercial and investment banking functions, making it illegal for commercial banks to also operate as investment banks.

The Clinton/Rubin/Goldman Sachs repeal of Glass-Steagall opened the way for commercial banks to once again get into the investment banking business. It also opened the door for Goldman Sachs to position itself as the most powerful player in the world economy, to the point where they control everything from the TARP bailout and the taxpayer-funded stimulus money to the terms of the long-term Treasury bonds that countries use to finance their sovereign debt.

After Clinton and Rubin engineered the repeal of Glass-Steagall and the distinction between commercial and investment banks disappeared once again, the stage was set for the crash of 2008 that got Obama elected.

The difference between the Great Crash of 1929 and the crash of 2008, which gave us current economic downturn, is that the 2008 meltdown was intentional.

We're now experiencing the second stage of that crash: the creation of an international financial power elite that is determined to control the world's economic activity and subjugate the people of this planet.

Because of the repeal of the Glass-Steagall Act, and because banks are now permitted to engage in high-risk brokerage activities, including selling insurance and underwriting securities, they're hedging their risk through such strategies as short-selling and continuing to develop and trade extremely risky financial derivatives.

Let me put it another way: We're being set up for an economic meltdown similar to the one that triggered the Great Depression, but this time it's going to occur on a global scale and it's unlikely that we'll be able to recover within even the next several decades.

Dodd-Frank: Aiding and Abetting the Enemy

With Glass-Steagall out of the way, new horizons opened up for the financial behemoths who sought to rule the global economy.

First, they had to get one of their own into the presidency.

I was one of the first to explain how they did it.

The repeal of Glass-Steagall enabled banks to engage in trading high-risk derivatives in order to hedge their exposure in the mortgage market, and the losses they racked up caused the financial crisis that ushered Obama into the White House. In order to prevent a repeat of the same scenario, Congress enacted the Wall Street Reform and Consumer Protection Act.[47]

The legislation is presumably written to prevent future government bail-outs of too-big-to-fail financial institutions.

Don't believe it.

That legislation—also known as the Dodd-Frank bill, after its sponsors, Connecticut Democratic senator Chris Dodd and Massachusetts Democratic congressman Barney Frank—was supposed to restore regulation to the financial industry in the wake of the 2008 crash.

The problem with the legislation starts with the bill's sponsors.

Chris Dodd is a member of the Bilderberg Group. He's also a member of the Council on Foreign Relations. Barney Frank is a member of the CFR. Which means these two are just the people the international financial elite were looking for to rubber-stamp their global financial takeover by removing any serious regulatory power from the legislation that was supposed to rein in the big banks.

Dodd-Frank does almost nothing to prevent another financial meltdown from happening. It waters down the separation of commercial and investment banking functions to such an extent that a repeat of the 2008 debacle is happening again before our eyes.

Although banks do face limits on how much of their capital they can invest in hedge funds or private equity funds (three percent), the bill allows banks to continue to engage in derivatives trading, requiring them to spin off only the riskiest of these trades. And the language of the 2,000-page bill is murky enough that banks will still be able to engage in trading derivatives tied to interest rate swaps. They're also allowed to continue to act as hedge funds for themselves, trading such assets as commodities and credit derivatives in order to minimize their risk.[48]

The Dodd-Frank bill was supposed to be about eliminating risky trading and requiring banks to manage their affairs responsibly and to bear the responsibility if they don't do so.

In other words, there would presumably be no too-big-to-fail banks under Dodd-Frank.

As I see it, Chris Dodd and Barney Frank did nothing to change things. All of the major banks are still too big to fail, they're still putting their clients' money at risk because they can still engage in hedging their positions, and you and I are still on the hook for their losses.

But don't just take my word for it. Here's what Neil Barofsky, formerly the TARP Inspector General, charged with overseeing how the bailout funds are managed, has to say about it:

> The continued existence of institutions that are "too big to fail"—an undeniable byproduct of former Secretary [Henry] Paulson and Secretary [Tim] Geithner's use of TARP to assure the markets that during a time of crisis that they would not let such institutions fail—is a recipe for disaster. These institutions and their leaders are incentivized to engage in precisely the sort of behavior that could trigger the next financial crisis, thus perpetuating a doomsday cycle of booms, busts, and bailouts.[49]

Tyranny of Fiat Money

You may not know it, but the money you spend to pay your mortgage and buy food for your family and gas for the car might as well be counterfeit.

When Richard Nixon took us off the gold standard in 1971, he ushered in the era of counterfeit money.

Economists don't use the term *counterfeit*. They call it "fiat currency."

Same thing.

Fiat money is currency that is not backed by a resource such as gold but that is valued based on its relative scarcity and on the faith the people have in the currency. Unlike money tied to a resource such as gold, there is no limit on the amount of fiat currency that can be created.[50]

Fiat currency is the basis of the financial manipulation bordering on fraud that the current administration is engaged in.

Thanks to a U.S. dollar that has no anchor in gold, combined with a Fed under Goldman Sachs alumnus Ben Bernanke and a Treasury Department run by Tim Geithner, another GS crony willing to do anything to keep American oligarchs' financial heads above water and their pockets lined, there is no limit to the lengths the Obama administration will go to perpetuate that financial chicanery.

The value of the dollar gradually became less and less connected to what was happening in the real economy, the marketplace where commodities, manufactured goods, and services are exchanged.

I've warned you about this over and over.

When the Fed and the Treasury conspired to print hundreds of billions of dollars in what they called Quantitative Easing, they accelerated the process of devaluing the dollar and the other financial instruments and processes that are linked to it.

I call it financial fraud.

Former Treasury secretary Alan Greenspan agrees with me. He just doesn't call it by the same name I do.

Here's how he puts it: "The United States can pay any debt that we have, because we can always print money." [51]

This is the same scenario that led to runaway inflation in post-World War I Germany, when it took a wheelbarrow full of German marks to purchase a loaf of bread. It's the same process that led the government of Zimbabwe to print a $100 trillion Zimbabwean dollar note. Because of the hyperinflation rampant in Zimbabwe, that note wouldn't even buy a loaf of bread. [52]

The threat of runaway inflation is beginning to surface in the U.S.

Our primary financial instruments and measures, from the dollar to the Treasury bond, have lost 65 percent of their purchasing power against natural resources. Economists and other financial analysts and commentators continue to focus on such things as GDP and corporate earnings, but the real issue is that the value of our fiat currency in terms of *real* purchasing power has diminished drastically in less than a decade. Paper currency is losing its purchasing power against natural resources, precisely because we no longer have a resource-based dollar.

The purchasing power of the dollar to buy life's necessities—including food, shelter, and energy—is the true measure of the dollar's worth. Food and energy costs are eating up a greater and greater percentage of Americans' earnings. In addition, as the value of the dollar collapses, the value of certain types of assets also collapses. Housing is suffering badly from rising real costs. It took a hit in the housing bubble; now it's having difficulty recovering its value. The largest and most important asset of most Americans is their home, and their homes have not recovered value against other necessities of life. [53]

The way I see it, the Obama administration is creating a permanent underclass who are rapidly moving toward third-world status, given the restraints on economic growth the Obama administration is instituting.

Americans' incomes have dropped to the same levels they were in 1996.[54]

The poverty rate has risen to its highest level in 50 years, with 46 million Americans (15 percent of the population) now living below the $22,000 poverty line.[55]

I maintain that inflation plays a big part in these statistics.

The Fed is saying just the opposite.

They're keeping interest rates low because in their intentionally myopic view "inflation isn't a long-term risk." They've been flooding the market with billions of increasingly worthless dollars as part of their strategy of quantitative easing, and that's driving up food and energy prices and eroding people's purchasing power. Nonetheless, they're going to keep interest rates "accommodative"—that is, "low." They should be raising rates in order to slow down inflation, but they won't even admit that inflation is occurring.[56]

It's not just happening here.

Inflation, as measured by the declining purchasing power of fiat currencies, is also rising dramatically around the world. The combination of low wages, rising taxation, and inflation have come together to reduce the real goods that fiat currencies, including the U.S. dollar, will buy to levels not seen in more than a century, according to one analyst.[57]

The U.S. Department of Agriculture report is an indicator of how expensive food has become. The squeeze is getting so bad that one in seven Americans, nearly 15 percent, are resorting to using food stamps to buy their daily bread, up from about seven percent only a few years ago.[58]

Precious metals are going in the same direction.

Gold is trading at or near all-time highs. Silver would be trading at all-time highs, but Comex (the commodities exchange) stepped in to stop silver's surge in order to bail out several key banks that were using silver as a hedge.

Here's something you won't hear from most financial analysts. The big financial institutions have accumulated so much power that in the space of four days in May 2011, silver lost about 25 percent of its value.[59]

Under the new post-Glass-Steagall banking rules, JPMorgan Chase had been hedging its risk by engaging in the short-selling of silver. The bank was already being sued for allegedly trying to drive the price of silver down in order to make huge profits on its short positions of the precious metal.[60] The problem was that the bank was using its own stock to collateralize its short silver positions, and when the price of an ounce of silver rose above the price of a share of JPMorgan Chase stock, the bank called in favors in order to reduce the price of silver.

Enter Comex, the exchange on which silver is traded. At JPMorgan Chase's bequest, Comex began raising the margins required for traders buying silver futures. This had the effect of causing the price of silver to drop significantly, saving JPMorgan Chase billions of dollars in its hedge against risk.

It's collusion at the highest levels of the financial industry.[61]

It's only one of many examples I've pointed out of how the big banks are manipulating markets in order to preserve their profits and those of their financial fellow travelers while people who buy commodities such as silver in the expectation that their value will increase are the ones who get hurt.

It's not just in the United States that it's happening.

Portugal, Ireland, Italy, Greece, and Spain (appropriately known as PIIGS) are being victimized by the same forces now operating in the United States. Their economies have been taken over by the global financial autocrats, and their governments are in the process of committing their citizens to pay back debt owed to the countries and financial institutions that are bailing them out, an amount of money that is far beyond what each of their economies is capable of generating over several lifetimes.[62]

Once again, it's the global crony capitalists, headed by Goldman Sachs, who are trying to manipulate the system for their own gain.

For years Goldman Sachs has played a lead role in brokering the sale of the sovereign bonds that Greece, Spain, France, and Italy use to finance their debt. At the same time they were underwriting European countries' sovereign bonds, the autocrats at Goldman Sachs were also brokering derivatives that bet against the very bonds they were selling to their clients.[63] Many European financial analysts were convinced that Goldman Sachs's underhanded dealings were instrumental in causing or contributing to the European debt crisis.[64]

Finally, after Goldman Sachs had helped undermine their bond sales and their economies, Europe banned Goldman Sachs—and other large

American banks—from participation in their bond sales. Arlene McCarthy, Vice Chair of the European Parliament's Economic and Monetary Affairs Committee, put it this way: "Governments do not have the confidence that the excessive risk-taking culture of the big Wall Street banks has changed and they still cannot be trusted to put the stability of the financial system before profit." [65]

Another commentator put it like this: "The French people would riot in the streets if we chose Goldman." [66]

The European debt crisis is contributing to U.S. economic uncertainty. Germany's representative to the European Central Bank (ECB) resigned on September 9, 2011, in protest against the bank's role in helping bail out European countries on the brink of bankruptcy, and the U.S. stock market fell more than 300 points in response. [67] The ECB is moving toward the same threat of collapse as U.S. banks reached in 2008 because it is heavily invested in the increasingly risky sovereign debt of countries like Portugal, Italy, Ireland, Greece, and Spain.

A similar thing is happening in the United States, where the Federal Reserve and other U.S. governmental agencies held $5.351 trillion in U.S. debt as of September, 2010, making our own government, at nearly 40 percent, the largest single holder of our debt. [68]

The Fed doesn't have enough money to purchase U.S. Treasuries, so it prints it, then buys U.S. Treasury bonds. As part of the second round of quantitative easing—called QE2—the Fed bought some $75 billion in Treasuries every month, artificially propping up their value.

The combination of quantitative easing, out-of-control spending, and debt growth had consequences. In April 2011, Standard & Poor's cut its outlook on our country's debt to "negative," down from "stable." S&P's revised outlook report said that in the next two years the ratings agency expected to have to downgrade Treasuries from triple-A status. [69]

The Obama administration's response?

It used the drop in the Dow Jones average to threaten lawmakers, not to do something to reduce the debt, but to raise the debt ceiling so the government could continue its unrestrained borrowing. [70]

It didn't take two years for S&P to downgrade our credit rating.

It barely took two months.

In August, 2011—*after Congress and the president had reached agreement on a debt ceiling deal*—S&P downgraded the U.S. credit rating from AAA to AA+ based on our rising debt burden and the political risk that the U.S. will not be able to get its unbridled spending under control.[71]

The Tools of Obama's Leninist Economy

In August, 2011, the U.S. economy created *zero* net jobs for the first time since the 1940s. More than 450,000 new government jobs have been created during the Obama administration, while the private sector has lost somewhere between three million and ten million jobs.[72]

Here's how I see what's happening in Obama's America: American workers are rapidly becoming the equivalent of tenant farmers in the rural South during the 1920s and '30s. More and more, they work the job that the government assigns to them, shop at the government store with money (welfare payments, food stamps) the government gives them for goods and services (Chevrolets and Chryslers and college educations and medical care) produced and delivered by the U.S. government.

Obama is determined to turn the U.S. into the next Cuba. Eighty percent of Cuba's workforce is employed by the Cuban government. Their pay? Twenty dollars a month. The benefits include free education, free health care, and cheap housing, transportation, and basic foods.[73]

As I've made clear, the people responsible aren't about to admit there's anything wrong with the economy and how their policies are turning America into the new Cuba. Tim Geithner will continue to treat us to disaster stories about what will happen if we don't raise the debt ceiling again.

The problem is that he won't tell us what's going to happen now that the debt ceiling has been raised.

I'm one of the only people doing that.

Monetary policy is not going to tighten after QE2. U.S. and European banks are not going to raise their interest rates. In fact, it's likely we'll see QE3 soon, despite the fact that all this fiat money is circling the globe and killing everybody's purchasing power because of skyrocketing food and energy prices.

The day after Congress passed the new debt ceiling legislation, China blasted the United States for its "madcap" borrowing, then began to unload some of the $1 trillion in U.S. Treasuries it holds.[74] Russian Prime Minister Vladimir Putin accused the U.S. of "leeching on the world economy."[75]

The earthquake/tsunami disaster in Japan means that that country, too, will likely have to begin to unload the more than $800 billion[76] in U.S. Treasuries it owns, or at least stop purchasing our debt as it tries to recover economically. And the upheaval in the Middle East means that countries in that region will be less likely to continue to finance our debt.

By December, 2011, because of rising concern over the growing economic instability among Eurozone nations, U.S. Treasuries had become more attractive to institutional investors. China remained the number one destination for our bonds at $1.13 trillion, while Japan had increased its holdings to $979 billion.[77]

Bottom line: Despite the fact that we're in somewhat better shape than Eurozone countries, we're not immune to the growing world economic crisis, and we're going to have to pay higher and higher interest costs on the money we borrow to finance our spending.

The U.S. economy hasn't been at this much risk of collapse since just before the Great Crash in 1929, and things won't be getting better anytime soon. Even though Barack Obama continues to try to sell his "recovery," we're three years into the Obama recession and things have not even begun to turn around.

I've lived through the four serious U.S. recessions since the end of World War II. In every one the economy recovered completely within three years, the same point we're at in the Obama recession.

Here are some numbers to compare.

In the previous recessions, our real gross domestic product was an average of 7.6 percent higher after three years than it had been when the recessions began. In the Obama recession, *real GDP is up one tenth of one percent* in that time.[78]

That's nothing more than a rounding error.

The annual income of every family of four in the United States should be more than $16,000 higher than it is today, 3.5 million more people should be employed today than currently have jobs.[79]

Instead, we've got Obama's Leninist version of a "recovery."

It's what happens when a martinet like Barack Obama teams up with financial pirates determined to highjack the global economy.

The best Obama has been able to come up with in defense of his economic policies was an address delivered on April 13, 2011. It turned out to be a collection of lies in defense of his economic policies. The normally restrained *Wall Street Journal* characterized it this way: "toxic," "ludicrous," "dishonest even by modern political standards." The speech was filled with "blistering partisanship and multiple distortions [of] the kind presidents usually outsource to some junior lieutenant."[80]

In Obama's chilling version of our totalitarian economic future, the government takes over a greater and greater share of the private economy, and, in league with Wall Street oligarchs and labor union thugs, dominates the industrial and financial landscape.

It's crony capitalism at its most destructive, designed to concentrate all the power in the hands of a few at the expense of the many.

Worse, it's economic suicide, perpetrated by a global financial oligarchy determined to control the world's economy through any means at their disposal.

We're living the results of a 50-year takeover by leftist elites determined to bring the economy under their control. It's not a political battle; it's a battle for the soul of America. Because of it, we may be living through the *final* economic depression. Unless we can find a way to stop the financial tyranny being imposed on us, it's likely to become permanent.

CHAPTER 5

Tyranny of Obama's Radical Accomplices

Red October, 2011: Obama's Communist Revolution

October, 2011, marked the emergence into the public eye of the second Bolshevik revolution, this one in the United States of America, egged on by our first communist president.

You've seen the indicators of this revolution—the Occupy Wall Street mob of young, indoctrinated anti-capitalist students and other malcontents, the gangs of union thugs trashing the Wisconsin capitol in the spring of 2011, the flash mobs of disaffected youths using cell phones and social media to organize raids to steal merchandise from American small businesses.

When I tell you that Obama supports all of these mob actions, that he wants to use them to hasten the overthrow of the American government, you'd better listen. Our way of life is in jeopardy.

I'll explain it to you.

First, the people who make up Barack Obama's inner circle are lifetime leftist agitators. From David Axelrod to Rahm Emanuel to George Soros, they're proponents of mob rule in order to speed up their plans for assuming power.

To hear them tell it, their target is "the rich." In fact, the rich have nothing to fear from this administration. When the administration comes out in support of the Occupy Wall Street mob against the banks, they're merely deflecting attention from their true purpose. In fact, it was Wall Street that put Obama in power in the first place, as I showed you in *Trickle Up Poverty.*

So why don't the people on Wall Street care that Obama is supporting the rabble as they demonstrate against the banks, against capitalism in general?

Because Obama's supposed "tax the rich" plan won't touch them. It's not aimed at the truly rich, the people with million-dollar-a-year salaries and perqs that would make you and me blush. It's aimed at the small businessmen who employ 15 or 250 or 700 people, the everyday American entrepreneurs who make a good living and provide jobs for Americans but who are anything but "rich."

Obama has made it clear that he hates the Tea Party.

Does that come as a surprise to you?

The Tea Party is the American middle class. Of course the president hates them. They're the bourgeoisie that every socialist hates. They're the people who work for a living, start companies, employ other people. They're the people who pay the taxes Obama squanders.

In other words, they're dangerous to Obama's revolution. Their economic success has got to be shut down.

Obama is targeting the middle class, so the big bankers could care less what he says about them or about the unwashed, unshaven, uneducated vermin who think they are promoting the downfall of the rich. His "tax the rich" scheme is aimed at people who are far from wealthy but who are nonetheless already taxed at a rate near 40 percent. It doesn't affect the truly rich, who are much more concerned with the capital gains tax. That's because they know enough not to take their money as income but in other forms that are not taxed as income.

Do you think Warren Buffett is worried about paying income tax?

Buffett, one of the richest men in America, earned nearly $63 million in 2010, of which about $40 million was taxable. He paid less than $7 million to the IRS, a 17.4 percent tax rate. He's advocating that people who make $1 million annually be taxed at the highest rate, and Obama is going to take that definition of "millionaire" down to people earning $200,000 annually.[1]

Obama was elected on the backs of hedge fund operators who made more than a trillion dollars betting against the U.S. economy in 2008. If we're looking for assets to seize, that's where we should start: with the hedge fund barons. In large part because they bet against the housing market in the crash before the 2008 election, American people lost their homes. The hedge fund operators gained their fortunes at the people's expense.

They haven't stopped ripping us off, either. The $2 trillion hedge fund industry outperformed the S&P 500 stock index for the first eight months of 2011, raking in more than $50 billion during that time. In August alone hedge funds gained $6.1 billion.[2]

If Obama is looking for a place to start an economic recovery, he should start by seizing the assets of these hedge fund vermin.

So what does this have to do with the rabble in the streets?

They don't have the slightest idea of what's going on.

They don't realize that they're nothing more than "useful idiots," being taken advantage of by Obama and Pelosi and Reid. The street mobs' usefulness comes because they deflect attention from the left-wing cabal that's going on right before our eyes. They're not attacking Obama and Pelosi and the other leftist politicos who are taking over the United States, and so the president and his henchmen let them demonstrate.

New York Magazine went into Zuccotti Park, where the New York City Occupy Wall Street demonstrators hung out, eating like kings and behaving like vermin. The magazine interviewed 50 Wall Street Occupiers to find out if they knew what they were talking about.

They didn't.

They didn't know what the Dodd-Frank Act was. As I explained to you in chapter 4, it was supposed to rein in the big banks and prevent them from engaging in the practices like marketing collateralized debt obligations and mortgage-backed securities that caused the financial meltdown of 2008 that led to Obama's election. It not only failed miserably to do what it was supposed to, it's very likely that it was never intended by the lawmakers who cobbled it together to do anything to slow down the spread of bad debt and the collapse of the economy.

The demonstrators didn't know that Ben Bernanke was the Fed chairman or that *SEC* stands for Securities and Exchange Commission. Two of the respondents said it stood for the Southeast Conference.

They had no idea about marginal tax rates. One of those answering explained that he knew "a whole f*cking family of billionaires who don't pay a thing."

The demonstrators overwhelmingly—94 percent—thought that more federal money was spent on defense than on health care, pensions, and education. In fact, less than half as much of taxpayers' money is spent on the military than on those three. But you wouldn't expect these freeloaders to know anything about paying taxes.[3]

The longer they're in the streets, capturing the headlines, getting the leftist anti-American message out there, the better it is for the subversives in the administration who are truly undermining the country as they seize more and more power over every aspect of our lives.

What you don't know is that the Russian Revolution of 1917—the one I compare our emerging revolution to—was not a revolution of the working classes. Karl Marx, the founder of communism, was the son of a lawyer, not a member of the proletariat. Friedrich Engels, his co-conspirator, was the son of a wealthy German cotton manufacturer. Vladimir Lenin, who led the revolution, came from a comfortable middle-class family.

The original communists were the hippie trust-fund cases of their time, and yet they were able to incite the poor and starving people of Russia into revolt.

Note, though, that there is no starvation in America. Most of the gutter rats you see demonstrating in the streets today are well-fed products of comfortable middle-class homes, too stupid to know they're being used to promote a new communist society that they think will make it so they don't have to work for a living like the parents who are still supporting them do.

You may not be aware that the "revolution" that's occurring today in the Occupy Wall Street movement has an American counterpart. Forty-two years ago, the Weathermen—an American communist revolutionary group—took to the streets of Chicago in a demonstration they called "Days of Rage" against the "imperialist oppressors." The purpose of the Days of Rage was to initiate the destruction of the American system, as it was outlined in a manifesto titled "The Elections Don't Mean Sh-t—Vote Where the Power Is—Our Power Is In The Street."

It didn't work out that way.

The demonstration was such a dismal failure that it caused the communist movement to change its strategy dramatically. The movement went underground, with radicals like Obama's friend and ghostwriter Bill Ayers throwing bombs from the shadows rather than demonstrating in the streets. The movement reemerged, its people now wearing business suits and skirts as they merged into the mainstream, planning a less direct revolution than the one they initially envisioned.

Professors like Frances Fox Piven authored what is known as the Cloward-Piven strategy for bringing down capitalism through the welfare system, as I explained to you in *Trickle Up Poverty*. Sure enough, Piven was one of the speakers at the Occupy Wall Street demonstrations. Piven railed against the bankers, not just as greedy thieves but as "cannibals . . . because they are eating their own—us!" That kind of rhetoric, so typical of the leftists of the 1960s, is reemerging today as the Obama administration looks on approvingly. It's the same Marxist propaganda that brought Russians to their knees nearly 100 years ago and led to 70 years of their being enslaved by the communist elite. It's the same political philosophy that killed 60 million Chinese under Mao's rule and that has led to the North Korean darkness today.

What you hear from the demonstrators themselves gives you an idea of just how ignorant they are: "I got warrants. I'm running from the law. I'm not even supposed to be here, but it's as good a spot as any to hide." "On my third day, they had smoked salmon with cream cheese. You know how much smoked salmon is a pound? Sixteen dollars. I eat better here than I do with my parents!"

General Electric CEO and Obama jobs czar Jeffrey Immelt—the guy whose company pays no corporate taxes and who is moving two of his businesses to Red China—chimed in with this about the demonstrators: "[W]e have to be empathetic and understand that people are not feeling great."[4]

Brian Phillips belongs to "NYC General Assembly," a group that has had a big part in the Occupy Wall Street demonstrations. He's clear about what he wants to see happen: "My political goal is to overthrow the government."[5]

Warren Goldstein, chairman of the Department of History at the University of Hartford, shouted to the demonstrators: "You are fulfilling the words of the prophet Isaiah. You have thrown off the yoke. Occupy, occupy, occupy!"[6]

The demonstrators, terminally obedient to the leftist professors who mentor them in the art of the free lunch, are cutting classes from their $50,000-a-year elite colleges to defecate on cop cars and sit around in circles praising the likes of the real terrorists their professors have taught them to emulate.

The Occupy Wall Street demonstrators have one important characteristic in common with Barack Obama and Adolf Hitler: They're blatantly anti-Semitic.

Bard College, which has loaned out many of its students to the Occupy Wall Street protest, has made the International Solidarity Movement an official campus organization. The ISM's stated purpose is to take action in order to destroy Israel. One of Bard's largest contributors is George Soros, who donated $60 million to the anti-Semitic hate mongers who run the school.[7]

Jewish Funds for Justice (JFFJ) is a communist-socialist group funded by George Soros. JFFJ staged Occupy Yom Kippur demonstrations, the Jewish equivalent of OWS. The group's communications director had this to say about the Occupy Wall Street demonstrations: "For many of us, social justice is where we find our Judaism." A JFFJ board member who teaches rabbinic literature at the American Jewish University praised the OWS movement in a piece he wrote for Soros's Center for American Progress. He talked about the anti-Semitic protesters as somehow representing a "just society."[8]

The words of the protesters themselves echo their mentors' anti-Semitism. In Los Angeles, one local schoolteacher said this: "I think that the Zionist Jews, who are running these big banks and our Federal Reserve, which is not run by the federal government . . . they need to be run out of this country." She lost her job as a result of her comments. Normally those on the left aren't punished for being racists, but public uproar was so strong that the teacher had to be fired. I'm certain she'll be hired back—under the radar—in some other position after the uproar dies down.

A New York protester had this to say: "[A] small ethnic group constitutes almost all of the hedge fund managers and bankers on Wall St. They are all Jewish. There is a conspiracy in this country where Jews control the media, finances. . . . They have pooled their money together in order to take control of America."

I expect that out of anyone who has been brainwashed in the U.S. educational system.

That includes Barack Obama.

Obama stood by, applauding the protesters, saying nothing against their blatant anti-Semitism. Jewish groups are speaking out against the president: The Emergency Committee for Israel released a video call for the president to take a stand against the demonstrators' anti-Jewish rhetoric and urging president Obama to take a stand against such hate-filled rhetoric.[9]

I've never understood how Jewish people could support this president and his anti-Semitic views. I call for the Jewish community to return to its senses and stand up against their own people who support this Israel-hating movement and our blatantly anti-Semitic president.

When he finally did comment, here's what the president had to say: "I think that it [the Occupy Wall Street movement] expresses the frustrations that the American people feel."

It's not a surprise to me that the president failed to speak out against the anti-Semitic demonstrators. When he was speaking at the Congressional Black Caucus awards banquet back in September 2011, he was comparing the tax rates paid by the rich and those who worked menial jobs for a living. He made what used to be known when I was growing up as a "Freudian slip." He said: "If asking a billionaire to pay the same tax rate as a Jew, uh, a janitor makes me a warrior for the working class, I . . . have no problem with that."[10]

That's not an accident. It reflects something very disturbing to me in the president's way of looking at the world. His own anti-Semitism is creeping in to the decisions he makes, the words he speaks.

The Government-Media Complex is complicit in this dangerous anti-Semitism.

CNN asked this question: "Would Jesus occupy Wall Street?"

Of course, when it's convenient for them, when turning to religions benefits their cause, the left is willing to invoke Jesus's name. They asked this question in order to make the point that Christians should support the rabble in the streets of U.S. cities.

The first problem with the assumptions underlying the question is this: Jesus was a Jew, and Jews aren't welcome among the "occupiers." Jesus would be pro-Israel.

He'd also be anti-communist. He instructed his followers to care for their families, their neighbors, the poor, and the sick. He did not say that we should turn our responsibilities over to the government.

When he confronted the moneychangers in the temple, he was not somehow striking out against banks and he was not doing so on behalf of the "masses." He was defending God's Temple, which the moneychangers had taken over.

Jesus would walk among the occupiers for one reason only: To point out to them that they were sinners, that rape, assault, bearing false witness, sloth, and coveting thy neighbor's possessions—all of which these occupiers are committing daily—are sins, and that they must be stopped.

He wouldn't have cared at all about the occupiers' political causes. In fact, he would have railed against having his name co-opted by such a vain, self-centered, destructive mob for their godless cause. The OWS mobs would have assaulted and spat on Jesus if he had walked among them.

For CNN to even imply otherwise is the height of hypocrisy.

The ranking Democrat in the U.S. House, Representative Nancy Pelosi, Socialist Party, California, followed the president's approval of the mob behavior with this: "I support the message." [11]

The president and the former speaker should be speaking out against what is really going on with this bunch of unwashed would-be radicals living off their parents. What they really want is to be able to live off the state for the rest of their lives.

The *Harvard Crimson* newspaper gave space to a group called the Revolutionary Communist Party, a fringe group that still holds out for the same things that Bill Ayers, who wrote Obama's book *Dreams from My Father*, wanted in the 1960s. Here is some of what this group stands for:

- Violent overthrow of the government.
- A separate region of the country for African-Americans (the South) and Hispanics (the Southwest). Special elections in which only African-Americans and Hispanics could vote would be held to decide the issue in the two regions. All people not of the appropriate race or ethnicity would be forced to relocate if they lived in the South or the Southwest.
- Prohibition of capitalism.
- Amnesty for all illegal aliens living in the United States, with the exception of "capitalist enemies of the state."

The Occupy Wall Street website lists dozens of "demands." Here are a few of them, so you can get a sense of how deranged these people are:

- Repeal the Taft-Hartley Act. Unionize ALL workers immediately.
- Raise the minimum wage immediately to $18/hr. Create a maximum wage of $90/hr to eliminate inequality.
- Institute a 6 hour workday, and 6 weeks of paid vacation.
- Institute a moratorium on all foreclosures and layoffs immediately.
- Repeal racist and xenophobic English-only laws.
- Open the borders to all immigrants, legal or illegal. Offer immediate, unconditional amnesty, to all undocumented residents of the US.
- Create a single-payer, universal health care system.
- Pass stricter campaign finance reform laws. Ban all private donations. All campaigns will receive equal funding, provided by the taxpayers.
- Institute a negative income tax, and tax the very rich at rates up to 90 percent.
- Pass far stricter environmental protection and animal rights laws.
- Ban the private ownership of land.
- Make homeschooling illegal. Religious fanatics use it to feed their children propaganda.[12]

Do you see what the OWS movement has devolved into?

Do you understand just how far this administration and its radical leftist labor union accomplices have unleashed on America?

Let me explain it to you.

More than 200 violent crimes—including rapes, knifings, and murders—have been committed at Occupy Wall Street sites around the nation.[13] We've witnessed public urination and defecation. The OWS sites quickly become open-air drug markets. Disease is rampant, including an outbreak of tuberculosis in Atlanta.[14] Public nudity, masturbation, and sexual intercourse are carried on daily by the vermin who populate these sites. They're characterized by incessant drumming bordering on psychological abuse by the aging children who have taken to the streets of our cities. Sanitary conditions are so dangerous in the festering mess that Occupy Wall Street protesters had created that after they were finally evicted the areas had to be decontaminated by sanitation workers in HazMat suits. Aggressive panhandlers mooch off anyone who dares get within a hundred yards of the demonstrators. Local small businesses like restaurants and coffee shops were forced to close their doors because the protesters drove away regular customers.[15]

That's what the rent-a-mobs that have been occupying public land in cities across the nation in the name of their anti-American, anti-capitalist, anti-morality, anti-Semitic causes.

They're walking billboards for the reason they are unemployable.

They're a combination of stupidity, laziness, and nihilism.

And each one of them borrows tens of thousands of dollars to learn these skills and values at Big Government U.

It wasn't until two months after the demonstrators had begun to occupy Zuccotti Park that the sniveling New York City mayor Michael Bloomberg finally decided to have his police department clear the demonstrators out. Showing incredible restraint, the police took care of business, finally removing the protesters and their tents and hosing down the area.

Within a few days, thousands of demonstrators showed up to participate in shutting down traffic and disrupting the business of New York City. Mealy-mouthed Mayor Bloomberg changed his tune, saying that the cops would respond appropriately if the demonstrators caused any trouble. Before it was over, nearly 300 were arrested as police managed to control the mob.[16]

That's what we're up against in the Occupy Wall Street movement—the most racist, violent, ignorant group of demonstrators since the 1960s.

If you think I'm exaggerating, that this is just the ranting of a small minority, then you haven't been paying attention. There's a reason Obama hasn't spoken out against these people. There's a reason Nancy Pelosi has given them her blessing.

During his campaign for the presidency, Obama promised to create "a civilian national security force that's just as powerful, just as strong, just as well-funded" as our military. His reason for doing this? It would help him advance his "objectives" for America.

No one in the Government-Media Complex bothered to ask him what he meant. But we're finding out now.

Teamsters union president James Hoffa Jr. explained what is going on in his Labor Day address—the same one Obama spoke at. Hoffa said, "President Obama, this is your army! We are ready to march! Let's take these son-of-a-bitches out!"

How is this a government-funded "civilian army"?

Let me explain.

Your taxpayer dollars fund the labor unions. You pay the salaries of public employee union members, and a portion of that money goes to the unions they belong to in the form of dues.

On top of that, Obama made sure that when the government commandeered General Motors and Chrysler, the unions took ownership of the companies, relegating stockholders to second-class status. The president has stacked the National Labor Relations Board in favor of the unions, and he's given unions exemptions from Obamacare in record numbers.

It's his way of funding his "civilian army." [17]

And that army is in the vanguard of the Occupy Wall Street movement.

The president and his anti-American cohorts are in fundamental agreement with the OWS demonstrators, and the longer their anti-freedom, anti-Semitic, anti-capitalist message is allowed to be broadcast to the American public, the more likely their agenda and their message become acceptable to the public.

Did you think that Occupy Wall Street was somehow a spontaneous uprising against banks?

It's exactly the opposite.

The Obama administration, through its shadow supporters, is using the spreading Occupy Wall Street demonstrations to get out their message. Ultimately they'll use it to clamp down further on our civil liberties.

Don't believe that?

George Soros's influence is everywhere. I told you about his anti-Semitic JFFJ, but his involvement doesn't end there. The demonstrations in New York and other American cities were started by a Canadian radical leftist group called Adbusters. The group's purpose is "to topple existing power structures." Adbusters used Soros money to stage a marketing campaign that was designed to bring the "Arab Spring" to cities in America. Soros's Open Society donated $3.5 million to the Tides Center, a San Francisco organization that distributes money to other leftist causes. The Tides Group gave $185,00 of that money to Adbusters.[18] The Working Families Party, along with the Service Employees International Union (SEIU) and what's left of ACORN, are also instrumental in organizing and spreading the demonstrations.

ACORN is still around—this time as New York Communities for Change (NYCC)—and they're still in the business of funding illicit activities in order to undermine the United States. They were revealed to be providing key organization services for the OWS movement. NYCC had been paying protesters to attend OWS demonstrations every day, despite NYCC's lying to the press, telling reporters they weren't. Insiders had outed NYCC, and the group scrambled to do damage control, shredding papers, destroying boxes of supplies with ACORN emblazoned all over them, threatening and firing employees who told the truth.[19]

Stephen Lerner, one of the commie rats who organized the demonstrations, spoke to the audience at Van Jones's Take Back the American Dream Conference in Washington, D.C., on October 3, 2011. In his speech he outlined the plans for expanding the demonstrations, including plans to "visit" millionaires' homes. That's the same thing the SEIU did in May, 2010, when 14 busloads of union mob thugs invaded the property of Bank of America general counsel Greg Baer and terrorized his teenage son in order to protest the activity of big banks.

The SEIU was active in the OWS movement, too. In mid-November, as the weather began to change, the 2.1-million member Service Employees International Union showed up at the Occupy D.C. encampment to help the demonstrators get settled in for winter. The AFL-CIO lets Occupy D.C. protesters take showers at a gym it owns near the demonstration site, and nurses have showed up to provide health care support.[20]

Other leftist groups are coming out of the woodwork to support the rabble in the streets.

Are you beginning to get the picture?

Are you starting to understand that this is only the first stage of a radical, potentially violent takeover of our country by Barack Obama and the forces of the left with whom he stands?

They're even using the phrase "Days of Rage" to describe the Occupy Wall Street demonstrations. They're also being compared to the "Arab Spring." Former 1960s leftist radical and ACORN founder Wade Rathke compared the two directly, using the rhetoric of Islamist radicals when he called the Occupy Wall Street demonstrations "anti-banking jihad." These radicals never went away, never changed their goals. They still want nothing less than the overthrow of our capitalist system.

Do you hear what I'm saying?

Do you realize that they're emulating their Nazi forebears with these tactics?

Germany was in economic turmoil after World War I. The communists were trying to take over the country, doing exactly what the Occupy Wall Street crowd was doing. The only difference is that the German communists were more violent. Out of the chaos and violence a group calling itself the Nazi Party emerged to battle the communists. Out of that struggle, Adolf Hitler came to power.

I am very worried that unless the authorities step in and put an end to these Occupy Wall Street demonstrations, we're going to see radical skinheads and Nazis come out of the woodwork and start street battles with these commie rats. This has the potential to lead to the emergence in America of a radical leader like Adolf Hitler.

It will also give an excuse to Barack Obama to impose martial law in America for the purpose of consolidating his power and that of the leftists and union thugs who are his allies in these demonstrations.

Nancy Pelosi, one of the most corrupt capitalists in Washington, doesn't seem to realize what she's saying. While I predict she will soon be coming up for investigation because she was instrumental in getting federal money directed to her brother-in-law as part of the "green energy" slush fund scandal, she's encouraging the Occupy Wall Street protesters, saying "God bless them!" [21]

Little do Obama, Pelosi, Emanuel, and friends know that the support of the rabble will soon backfire on them, as it did in the French Revolution. Soon the attackers will be attacked. Once heads start to roll, it's hard to stop them.

I predict that Red October in the White House will soon come to mark the beginning of the end for the most subversive administration in American history.

Mob Rule

The Occupy Wall Street "protests" are nothing more than an extension of the tactic of choice for the American left for most of the past century: mob rule.

Here are a few examples.

I've already told you about the 14 busloads of SEIU union members who terrorized Bank of America lawyer Greg Baer's teenage son in order to protest large U.S. banks.

Were you aware of this one?

The National Air and Space Museum in Washington was closed down on Saturday, October 8, 2011, by antiwar demonstrators who were protesting an exhibit about drone aircraft—the same aircraft the Obama administration these demonstrators support used to illegally kill American citizen Anwar al-Awlaki. [22]

The unions also unleashed a mob in the port of Longview, Washington. More than 500 longshoremen broke into a guard shack and took security guards hostage. They proceeded to cut brake lines on trains in the terminal and dumped grain waiting to be loaded and shipped. Longview police chief Jim Duscha said, "A lot of protesters were telling us this is only the start." [23]

The Government-Media Complex didn't bother to report the incident.

Barack Obama was silent, refusing to criticize these union terrorists.

They're even using the mob to beef up the president's employment numbers: Almost half of the 103,000 "new" jobs in the September jobs report came because striking union workers at Verizon came back to work.[24]

The leftist creeps who support Barack Obama are taking their cue from the labor unions and the totalitarian socialist ideologues of the 1930s. They're targeting middle America in an ongoing assault on our freedom that we haven't seen the likes of in years.

I've seen it coming for a long time, but now it's right out in the open.

When they're not physically assaulting their fellow citizens and destroying property, they're trying to provoke street war with Americans through abusive and provocative language. On Labor Day 2011, Obama vowed again to stand with unionized state employees, the same group that resorted to mob rule—trashing the Wisconsin capitol to protest the policies of elected officials, showing up at the houses of government employees whose policies they disagree with in order to intimidate their families.

Let me remind you about what happened in Wisconsin.

In November 2010, Scott Walker won the election for governor of Wisconsin, and his Republican Party gained a 19–14 majority in the previously Democratic-controlled state senate. One of his first acts as governor was to introduce legislation requiring Wisconsin public employees to pay part of their retirement and health-care benefits costs. The new law also limited collective bargaining rights for those employees.

Obama stuck his nose in the state's business the day after the legislation was introduced. With all the civility of a 1930s union thug, the president called Governor Walker's attempt to bring Wisconsin's budget under control by introducing sanity back into the budget negotiations "an assault on unions."[25] His administration then helped organize the union protests that effectively shut down Wisconsin's state government for weeks as Democratic state senators fled to a nearby state with their tails between their legs, illegally short-circuiting the democratic process.

At the same time, many of the state's teachers staged a strike. They called in sick in such large numbers that all the schools in the state capital of Madison and many others cities around the state were closed down. Many of the "sick" teachers, along with the students they were supposed to be teaching, showed up at the capitol to protest the fact that they actually had to contribute to their health-care and retirement benefits. Doctors who supported the teachers' position followed them there, writing out medical excuses for the teachers who were staging an illegal work stoppage at the expense of the children and Wisconsin taxpayers.

They were actually doing their students a favor by getting them out of the classroom. In Milwaukee's public schools, the teachers are doing a typically lousy job: Black students in that district have the lowest graduation rate of any school district in the nation.[26]

It wasn't long before AFL-CIO president Richard Trumka showed up in Madison to shout at his followers as he railed against Walker's initiative. Trumka didn't call Walker a Nazi, but many of the liberals in the crowd he addressed carried signs that did.

Trumka is the president of the AFL-CIO. He officially joined forces with Obama in a Big Labor-Big Government alliance on Labor Day 2010, in Milwaukee. One blogger described the event as "a midterm shotgun marriage of Beltway brass knuckles and Big Labor brawn."[27] Trumka is also the former president of the United Mine Workers of America (UMWA). In 1993, Trumka advised striking miners to "kick the s– out of" anyone who disagreed with what the UMWA was demanding from management.

Trumka had replaced the SEIU's Andy Stern as the White House union insider. In Trumka's words, "I'm at the White House a couple times a week . . . two, three times a week. I have a conversation every day with someone in the White House or in the administration. Every day."[28]

Trumka joined with another group, Organizing for America (OfA), in making sure the voices of the people of Wisconsin were drowned out by union organizers and other leftists. OfA is an arm of the Democratic National Committee that supports Obama's "agenda for change."

How closely aligned is Obama with these thugs? You can sign up to participate in a leftist mob at the website BarackObama.com. Just enter your zip code and the website will find a mob you can join in order to protest democracy.

The president is not about to abandon an arrangement that sees taxpayer money pay public worker salaries, which fund union dues, which then come back to Obama and other Democrats in the form of campaign donations. It's a blatant conflict of interest when Democrats "negotiate" with the very unions whose members pay their salaries and donate to their campaigns.

They're negotiating with their friends. They're not trying to manage taxpayer money, they're giving it away in the form of highly generous salaries—teachers work about 180 days a year for full-time pay—and lavish benefits packages that have helped push state and federal budgets into bankruptcy.

Democrats care nothing about taxpayers except that they're an endless stream of revenue that can be channeled to promote liberal interests and bribe liberal public employees into voting a Democratic ticket.

Obama's ties to his union cronies represent a direct conflict of interest for him where his constitutional duties as president of the United States are concerned. He consistently goes against the will and the best interests of the American people, and his anti-American agenda has got to be revealed. Events like the Wisconsin public employees dispute bring the issue and the deep corruption of the Obama administration into focus. The contrast between the thuggish language of Obama himself and his confederate Richard Trumka, compared with the even-tempered, rational Wisconsin Governor, Scott Walker, reveals just how rabidly this administration and the unions hate middle America, and the lengths they will go to in order to reduce our say in how our government conducts the people's business.

What happened in Wisconsin is being played out again and again in the increased union activity that Obama is drumming up as part of the 2012 election campaign. That election is ultimately about whether president Barack Obama will be allowed by the American people to continue his illegal and unconstitutional takeover of our nation through an unchecked expansion of government, or whether we the people will take back control from a leftist chief executive who is determined that we shall no longer have the right to govern ourselves but will be held captive by a government that is assuming control over every area of our lives, from personal and medical decisions to how our money is spent to what kinds of cars we can drive and what kinds of lightbulbs we can use.

What's going to happen if we let the administration and the union thugs continue to cause chaos in the streets of America as we move toward the election of 2012?

Let me make it clear to you just how high the stakes are in this fight.

We've been asking "Where are the moderate Muslims?" Now we have to ask, "Where are the moderate labor union members?" Moderate labor union members—like moderate Muslims—are too intimidated to speak out against the unions' policies of protecting bad teachers through tenure laws. Teachers know what will happen if they speak out against thuggish behavior of unions, just as Muslims know what will happen if they speak out against radical Islam.

As I see it, the Obama administration is raising the level of the rhetoric and intimidation to incite the people, to encourage them to take part in mob warfare that requires that the administration step in, possibly to declare martial law.

The parallels to the Nazis' rise to power are ominous. What's happening with Obama and his radical accomplices—especially the labor unions—is very similar to what Hitler's Brownshirts did as the Nazis rose to power.

They don't stop there, though. The out-of-control spending, runaway inflation, communist thugs inciting the people—these all echo what was happening in Germany in the 1930s.

I can't see Obama letting this crisis, which his own government has created, go to waste.

I'm convinced we're witnessing, for the first time in our history, a president of the United States take a page from the Nazi playbook as he tries desperately to consolidate his power.

The Tea Party is one of Obama's prime targets. I am sure that the Tea Party has already been infiltrated by leftists who will pretend to be Tea Party members and speak out as if they are, saying racist and violent things in the name of the Tea Party to give the administration a reason to step in and shut the Tea Party down. Tea Party members—and other conservatives who see the dangers of this administration—must take extreme care not to react to the insults and challenges that are hurled at them by Obama and his leftist cohorts.

It's one of the tactics developed by the godfather of the community organizing in America, Saul Alinsky. Barack Obama and many of his closest advisors—including Secretary of State Hillary Clinton and presidential advisors David Axelrod and Rahm Emanuel—are Alinsky disciples. Obama himself cut his teeth as a community organizer in Alinsky's hometown.

I want to make you aware of what this background means for those of us Obama is trying to subjugate.

Rules for Radicals

The president and his advisors and labor union leaders are disciples of Chicago-born Saul Alinsky. Alinsky was a career leftist political agitator. He literally wrote the book on how to be an effective community organizer: *Rules for Radicals: A Pragmatic Primer for Realistic Radicals.* The first thing you notice when you open the book is that it's dedicated to the devil: "[T]he first radical known to man who rebelled against the establishment and did it so effectively that he at least won his own kingdom—Lucifer."

Alinsky's earned his nickname—the father of community organizing—during a nearly 40-year career going around the country and stirring up "the masses" against their capitalist oppressors. He was the ideological father of Barack Obama, the community organizer we elected president. Obama spent time in that role in Chicago on Alinsky initiatives like Project Vote and the Developing Communities Project.[29] Here's what he said about the experience: It was "the best education I ever had, better than anything I got at Harvard."[30]

Obama isn't the only one in his administration who cut his teeth on Alinsky's radical teachings.

The key figure in the administration's push to bring America under leftist rule is senior advisor David Axelrod. Axelrod was mentored by Alinsky disciple David S. Canter. Canter founded the Chicago-based leftist journal *Hyde Park Kenwood Voices* and "helped educate [Axelrod] politically."[31] When leftists use the word "educate" they mean "indoctrinate." A piece about Axelrod in the Chicago Tribune had this to say: "[I]n his early years as a political consultant, Axelrod, following in the footsteps of his mentor, the political strategist Don Rose, carved out a reputation for himself as a skillful specialist working for local progressive candidates."[32] Rose was a member of a Communist Party front organization called the Alliance to End Repression.

Axelrod's parents were leftists, too. His mother, Myril Axelrod, was a journalist who worked for the commie tabloid newspaper *PM* in the 1940s. The journal was aggressively pro-communist and boasted among its writers communist sympathizer I. F. Stone, as well as a former editor of the communist periodical *Daily Worker* and a leader of the Communist Youth League.

Axelrod never tried to hide his communist sympathies. In an interview in which he explained that it might be necessary to keep the Bush tax cuts for the rich in order to extend them to the poor and middle class, Axelrod said, "We have to deal with the world as we find it, the world of what it takes to get this done."

Axelrod was quoting directly from *Rules for Radicals*.[33]

Rules for Radicals lays out Alinsky's strategy for promoting class warfare—the same thing Obama did when he toured the country promoting his 2011 "jobs" bill and trying to sell his "tax the rich" strategy. Alinsky's intention was nothing short of ending capitalism. In an interview he gave just before his death in 1972, Alinsky described our government as "American fascism." What is the only alternative? "Radical social change."[34]

In his 2008 speech accepting the Democratic nomination for president, Obama put forward his version of this Alinsky mantra: "Change we can believe in."

Alinsky and Obama both had the same outcomes in mind.

Obama and Axelrod aren't the only Alinsky followers in the administration. Hillary Clinton's senior honors thesis at Wellesley College was titled "THERE IS ONLY THE FIGHT—An Analysis of the Alinsky Model." Clinton's thesis was finally been made public against her wishes.

From the 1940s through the late '60s, Alinsky traveled the country staging sit-ins and other types of demonstrations, as well as boycotts. He worked to organize urban neighborhoods by politicizing their residents and getting them to unite against the "enemy." During that time, Alinsky carried a business card that said "Have trouble, will travel."[35]

Here's the most important thing you need to know about Alinsky: He put the political philosophy of the Obama administration down on paper.

First, Alinsky likened leftist radicals like himself to the people who stood up against Hitler in World War II. The "haves" that he was trying to overthrow were the equivalent of the Nazis. The "haves" still are Nazis for the Obama administration.

Alinsky advocated using minorities to expand the power of the left. Like Obama, he didn't care at all about making minorities' lives better, only about gaining power. Like all leftists—Obama included—Alinsky saw people not as individuals but as members of racial or ethnic groups.

Alinsky defined his agenda based on two things: "power" and "compromise." In his words, "Every organization known to man, from government down, has had only one reason for being—that is, organization for power in order to put into practice or promote its common purpose."[36] Alinsky's intent was "to organize for power: how to get it and how to use it."[37]

If that sounds familiar, it's because it's exactly what Obama bases his political philosophy on. The ends justify the means if the ends are achieving power.

During the health-care "debate," without a single Republican vote, against the will of the American people who protested in town hall meetings through the summer of 2009, and through a series of payoffs of hesitant Democrats in order to get even the Democratic votes needed for passage, the Obama administration defined itself as willing to engage in any behavior in order to achieve its ends.

It's classic Alinsky.

Another Alinsky principle: If you lose power, as Democrats did in the 2010 elections, compromise: "[T]o the organizer, compromise is a key and beautiful word. . . . If you start with nothing, demand 100 per cent, then compromise for 30 per cent, you're 30 per cent ahead."[38]

Radical Islamists have been paying attention to the Alinsky playbook, too. To them Americans' sense of right and wrong is a sign of weakness. Taking advantage of the West's sense of "fairness," Islamists in the Netherlands have managed to get the government to require that the name of Christ be spelled with a lowercase *c* because Christianity is offensive to Muslims. In Britain, images of Winnie the Pooh have been banned so that Muslims don't have to deal with pictures of a pig, which would be offensive to their sensibilities.[39] The practice of demanding that democracies bow to the will of the true enemies of freedom is an Alinsky tactic that every antidemocratic group from radical Islamists to the ACLU to the Obama administration takes advantage of at every turn.

Alinsky used personal attacks as another powerful weapon. The last of the 13 rules he laid down in *Rules for Radicals* is this: "Pick the target. Freeze it, personalize it, and polarize it." He explained that "[r]idicule is man's most potent weapon." The leftist media's vicious attack on Sarah Palin and their attempts to blame the murderous Tucson rampage by madman Jared Loughner on conservative hate speech represent Alinsky's tactics at their ugliest and lowest.

I should know. Chris Matthews, the leftist host of MSNBC's *Hardball*, blamed me for creating a climate of hate that led to Loughner's shooting spree, saying, "People like Michael Savage, for instance, who every time you listen to them are furious, furious at the left, with anger that just builds and builds in their voice . . . every night with ugly talk, ugly sounding talk."[40]

Alinsky's tactics are so much a part of leftist thinking that they've become second nature. Alinsky defined and put into practice every one of the principles and tactics that are in use by the Obama administration in America today. In his days as a "community organizer" Obama absorbed much of Alinsky's teachings. He's said so himself. As president, he's putting Alinsky's tactics into practice.

We're living under a government headed and run by vermin who see America as the enemy.

Don't believe me?

Here's how Alinsky's son explained how the Obama team put his father's teachings into practice at the 2008 Democratic National Convention:

> All the elements were present: the individual stories told by real people of their situation and hardships, the packed-to-the rafters crowd, the crowd's chanting of key phrases and names, the action on the spot of texting and phoning to show instant support and commitment to jump into the political battle, the rallying selections of music, the setting of the agenda by the power people. . . . The Democratic National Convention had all the elements of the perfectly organized event, Saul Alinsky style. . . . Barack Obama's training in Chicago by the great community organizers is showing its effectiveness.[41]

It was Alinsky who taught his followers to "never let a serious crisis go to waste." Those were the same words used by Obama's then Chief of Staff Rahm Emanuel.

The Obama administration takes his advice.

They used the underwear bomber crisis to make every airline passenger submit to invasive and dangerous X-rays or pat-downs every time they fly. It's another incursion on our rights and our privacy that marks a further step toward this administration's using a crisis, even a manufactured one, in order to advance its agenda of total control.

The United States is fast becoming Alinsky's America, with labor unions and crony capitalists and race- and agenda-based groups of thugs of every sort vying for a piece of the action.

The Obama administration exists for one purpose only: to amass the power to push through its dictatorial agenda and gain control for a group of blue-collar union bosses and white-collar financial industry cohorts who have ingratiated themselves with the president. And he owes it all to Saul Alinsky and the new generation of subversive community organizers and activists and radical leftist Democrats.

Rules for Beating Radicals

Here's some advice for you as you watch this battle unfold.

Never believe a word that comes out of a Democrat's mouth. Lying is their standard mode of operation, and they're not about to change now. They can't let us hear what they really believe, because most Americans would immediately call for their arrest on the grounds that what they say constitutes criminal activity.

Never negotiate with a leftist Democrat. There's no such thing as "good faith" where leftist liberals are concerned. I know. I've been there. I'm banned from entering Britain for being a passionate nationalist by corrupt British leftists of the same type as now preside in America. We've seen during the first two years of the Obama administration how the left operates when they have control over both houses of Congress and the executive branch. When we take back the presidency and the Senate in 2012, we've got to remember that our first job is not to work with Democrats and give them a chance to continue to influence America's future, but to do anything we must to begin to undo the damage they've caused.

Never compromise with a Democrat. As I've explained, for those on the left, any compromise is a victory. For real Americans, any compromise with the left is a loss. To allow even the smallest component of the leftist agenda to be considered legitimate is to grant a foothold that you may never be able to recover. When Democrats uses the word *compromise*, they mean that their enemies have given in.

Attack Democrats before they attack you. Democrats like our president have no core principles except the lust for power. Following Alinsky, they say, "Make the enemy live up to his/her own book of rules." Because those on the left are unprincipled but will nonetheless attack you if you violate one of *your* own principles, it is important to call them out based on your principles and demand that they act in a principled manner.

Obstruct Democrats at every opportunity. If a Democrat calls you an obstructionist, take it as a compliment. It's the Democrats who are the true obstructionists; they're the ones who stand in the way of individual Americans exercising their right to take care of themselves. They're the ones who want to make every decision for us, to obstruct our right to live our lives as we see fit. They're the ones who confiscate our money and squander it on an educational system that is nothing more than an excuse for teachers' union members to live off the public while undermining through negligence and self-interest the lives of the very students they're charged with educating.

Use the left's treatment of racial and ethnic minorities against them. Following Alinsky, the left uses minorities for only one reason: to help solidify its power. At the same time, it continues to work to make sure that minorities are held down. In fact, if minorities were to be given their freedom from Democratic oppression and encouraged to freely exercise their rights and bring their dreams into realization, the Democratic Party would simply disappear. It would have no group on whose shoulders it could climb to power.

We're in the fight of our lives for the survival of our nation. The policies of the left, including many Democrats now in positions of power in our government, are designed to do nothing less than eliminate the principles articulated in the U.S. Constitution as the foundation of our legal system by abrogating the rights of the individual and replacing them with the authority of the state to intrude into every aspect of our lives and determine how we must live, speak, and behave. What the Obama administration is attempting to do represents the most insidious abrogation of human rights in our nation's history. We have to stand up against Obama and the people who want to Occupy the United States, before it is too late.

CHAPTER 6

Tyranny of Obama's Corporate Cronies

One of the things I have learned watching the *Godfather* movies is that every tyrant needs accomplices, a band of brothers to carry out his dark designs. Running a mafia is simply too big of a job for one man alone; even a talented type-A overachiever can't do it by himself. Michael Corleone needs his consigliere and his henchmen.

So Barack Obama needs his merry band to run the alphabet soup of bureaucracies from the Agricultural Research Service to the Veterans Health Administration. And as president, he can appoint many of them. The president has the constitutional authority to appoint up to 3,000 senior government officials and put thousands more into positions like those of special envoys, special assistants, commissioners, task-force members, and members of other unelected bodies.

Czars Belong in Russia

Not that that was enough for Obama. In typical tyrannical fashion, the president immediately went beyond the bounds of the Constitution and started appointing czars, unvetted cronies chosen to do the left's dirty work without congressional approval. He brought in more than 40 czars, with unpublished salaries and secret agendas.[1]

I could write an entire book just about the often criminal and unconstitutional activities these guys engage in. They were appointed extralegally, and almost all of them are convinced that they're above the laws that you and I have to obey.

Take the case of Obama's environment czar, Carol Browner. She claims to love the environment, but the only things she really loves are regulations that don't apply to her. I did some checking into her background and I found something the major media wouldn't dare to report. When she was nominated, but not yet confirmed as head of President Clinton's Environmental Protection Agency (EPA), she wanted to cut down a 120-foot-tall oak tree in order to widen her driveway. But she lived in Takoma Park, Maryland, one of the most liberal bastions on the East Coast.

Takoma Park has one of the toughest tree protection laws in the United States. Even putting a flagstone path near a tree can result in a $1,000-dollar fine, as Takoma Park residents S. Craig Alexander and Amy Wasserstrom learned in August, 2009. While the fine was later reduced to $250 plus court fees and a scolding from the presiding judge, they got off lightly for their indiscretion. In fact, putting anything on the ground *within 50 feet of a tree* in Takoma Park requires a tree-impact statement.[2] Cutting down trees is strictly forbidden.

Takoma Park is legendary when it comes to enforcing its strict tree laws. Even TreeWorld.info, a website devoted to tree regulation around the world, notes how strict Takoma Park is: "You may want to do the right thing in this place because the city is proactive at busting people."[3]

What's more, Takoma Park employs an arborist named Todd Bolton to roam the small city looking for the tiniest of violations.[4] Repaving a driveway within 50 feet of a tree requires permission from Bolton. Even pruning a tree requires an almost impossible-to-get permit from the city. So Browner knew there was no way she was going to get a permit to cut down a beautiful, healthy oak that had spread its branches for more than 80 years, giving shade to passersby and a pleasant green glow to Browner's neighbors.

So she cut it down in the middle of the night, under cover of darkness. Or she had someone cut it down for her. The one thing we know for sure is that the 120-foot tree disappeared one night, leaving only a stump and a sprinkle of sawdust.

Of course, one of Browner's neighbors turned her in. She had to appear before the city council on more than one occasion, but eventually she was given an after-the-fact permit.

Neighbors say it is the only time they have ever heard of anyone in Takoma Park getting permission to fell a large, live, healthy oak.

Browner is typical of Obama's accomplices: She doesn't believe the laws and regulations apply to her—but only to the "little people."

Years later, Browner was forced to admit to being a tree killer in a little-noticed online discussion with the *Washington Post*: "Before I was at EPA, my husband and I did take down one tree that was *threatening* the sidewalk, driveway, other trees in area." [5]

Did you get that?

The tree was *threatening* the sidewalk!

And the fact that she planted some three-foot saplings to "replace" the downed oak doesn't make it right. She destroyed a 120-foot giant, whose branches saw Henry Ford's Model T's rumble underneath. While the Takoma law might be wacky, the intent of it is clear: to preserve the city's oldest residents, its trees.

But Browner got a pass for the liberal-run city. They didn't want to hurt one of their own.

Kenneth Feinberg, the so-called "pay czar," wants to set the salaries and bonuses of bankers. Sure, some of these bankers didn't do much except ruin the economy and buy expensive purses for their wives to put their dogs in. And write campaign checks to Obama. But I'm worried about the principle and the precedent.

If a banker is paid too much, the only people he is hurting are the people dumb enough to buy shares in his bailed-out bank. Let's not forget the principle at stake here: Once the government begins telling banks how much they can pay people in private industry, it won't be long before it starts telling building contractors, business owners, and manufacturing companies how much to pay their employees. Once the government thinks they have the right to do it, they will do it everywhere.

Now I come to the precedent. If anyone objects to the "pay czar" setting private industry salaries, that unelected czar can point to the banks and say the legal precedent is already established. Every office has a guy who works hard and another guy who is just coasting. You know what I mean. Do you want a pay czar—who's never met either one of these guys—saying they should make the exact same amount, no matter how hard they work? That's how the Obama tyranny works.

I've got something to say about Cass Sunstein, Obama's regulatory czar, in the next chapter, but I want to introduce him to you now. You think you know all about him? Well, I uncovered some new material that will outrage you and any thinking American.

This unelected panjandrum is supposed to supervise all the regulations issued by the federal government. It's a big job. The federal agencies put out several hundred thousand pages of new regulations every year. The new rules—and just the new rules, not the old ones—take up 11 times more pages than the New York and Los Angeles phone books put together.

And it is a complex job. Regulations are not laws, but rules based on laws. They are supposed to fill in the blanks in laws. And laws, which can only be passed by Congress, are supposed to be based on the Constitution. So you'd think that Cass Sunstein would be a big defender of the Constitution.

But you would be wrong.

Sunstein doesn't like the First Amendment to the U.S. Constitution, which protects free speech. He actually wrote a book called *Democracy and the Problem of Free Speech*. In that book, he says that many forms of speech should not be protected at all, including commercial speech—advertising, product presentations, billboards, and so on. He's also against speech that invades privacy—no more embarrassing stories about politicians and movie stars—and what he calls "hate speech," which amounts to criticizing anything that liberals don't want criticized. In the book he calls for a "large-scale reassessment of the appropriate role of the First Amendment in the democratic process."[6]

But the form of speech that Cass Sunstein dislikes the most is political attack ads on television and radio—the very ads that go around the mainstream media and educate voters about the records of liberal politicians. Of course, if politicians supported what the majority wants, there would be no attack ads. But when politicians support something that five or ten percent of special interests want against the will of 80 to 90 percent of the public, those political positions make politicians unpopular. And politicians don't want the public to know anything that would make them unpopular.

When Congress was debating campaign finance laws that would forbid people from putting political signs on their lawns fewer than 60 days before an election and banning people from spending money to buy independent political ads, Sunstein came to the defense of people who wanted to ban free speech. He said, "A legislative effort to regulate broadcasting in the interest of democratic principles should not be seen as an abridgment of the free speech guarantee."[7]

It shouldn't?

So he thinks that this kind of "hate speech" should not be protected by the First Amendment and that all political speech should be regulated by campaign finance laws. He likes the idea of banning political ads by independent groups 60 days before any federal election—the very point in time when voters start to get curious about the political records of candidates.

Instead he calls for a "New Deal" on speech that would give free air time for political candidates, strictly regulate what can be said about political issues, and deny access to the airwaves by citizen groups or principled individuals.[8]

And it is not just Obama's czars who flout the law and the Constitution. At least two of his cabinet secretaries have trouble paying their taxes.

None of these czars are poorly paid. While most sources list "salary unknown" for positions such as the Great Lakes czar, most czars earn north of $100,000, and some, like intelligence czar Dennis Blair, earn as much as $197,700. Poor Joshua DuBois, the faith-based czar who acts as a liaison between faith-based and secular community groups, couldn't break the 100K mark, earning just $98,000 a year. But he is the exception. Then there is the "pay czar," who is widely reported to have worked as an unpaid government employee. Turns out the pay czar should have been called the Pretty Well Paid Czar. He received a salary of $120,830 annually according to U.S. Treasury Department and U.S. Office of Personnel Management documents obtained by Judicial Watch under a Freedom of Information Act request.

When the news broke, his spokesman told the *Washington Examiner* that Feinberg received paychecks "every two weeks, but as soon as he got it, he endorsed it over and returned it." Yeah, right. Judicial Watch tried to get clarification from both Feinberg and Treasury officials, but neither would comment. Thus, said the *Examiner*, "[a]s a result, it could not be learned if Feinberg has been cashing the government paychecks, returning them to the Treasury Department, forwarding them to a charity, or otherwise not accepting them." [9]

Congress has tried to rein in Obama's unconstitutional czars. House Speaker Boehner, Senate Majority Leader Harry Reid, and President Obama struck a budget deal back in April 2011 for funding the remainder of fiscal 2011 that eliminated funding for four czars: Health Care, Climate Change, Auto Industry/Manufacturing Policy, and Urban Affairs. [10] Eager to blunt the Republican victory, the White House told Politico.com that three of the four positions were already vacant anyway: "Over the last several months, the White House has undergone a reorganization that involved the consolidation of several offices and positions. Included in that larger reorganization, earlier this year the Domestic Policy Council assumed responsibility for health care, as well as energy and climate change policy coordination and development in the White House. The agreements reflect those changes." [11]

The "agreements" didn't stay agreed to for long. Three days later, Obama issued a "signing statement," a written comment printed with a bill when a president signs it into law that makes some claim about the legislation's constitutionality and states whether the administration plans to ignore the legislation or implement it in ways they believe are best. In the signing statement he attached to the spending bill, Obama claimed that the passage about defunding the czars had a meaning *opposite* of what the words actually say. He wrote: "Legislative efforts that significantly impede the President's ability to exercise his supervisory and coordinating authorities or to obtain the views of the appropriate senior advisers violate the separation of powers by undermining the President's ability to exercise his constitutional responsibilities and take care that the laws be faithfully executed. Therefore, the executive branch will construe section 2262 not to abrogate these Presidential prerogatives." [12]

The bottom line: Obama will continue to pay his czars claiming his tyrannical presidential prerogatives. So much for his "agreement."

The prerogatives Obama claims are expressly forbidden by the U.S. Constitution: "The President . . . shall nominate, and by and with the advice and Consent of the Senate, shall appoint Ambassadors, other public Ministers and Consuls, Judges of the supreme Court, and all other Officers of the United States, whose Appointments are not herein otherwise provided for, and which shall be established by Law." [13]

Writing in the Federalist Papers on this very subject, Alexander Hamilton, President Washington's first secretary of the Treasury, said this in Federalist 69: "The President is to nominate, and, *with the advice and consent of the Senate* [italics in original], to appoint ambassadors and other public ministers, judges of the Supreme Court, and in general all officers of the United States established by law, and whose appointments are not otherwise provided for by the Constitution." [14]

Hamilton contrasted the checks and balances placed on U.S. presidents with the prerogatives of the tyrannical kings of England: "The king of Great Britain is emphatically and truly styled the fountain of honor. He not only appoints to all offices, but can create offices. He can confer titles of nobility at pleasure, and has the disposal of an immense number of church preferments. There is evidently a great inferiority in the power of the President, in this particular, to that of the British king." [15]

President Obama has rectified this "inferiority of power." He now claims the prerogatives of kings.

Obama's Cabinet: Tax Cheats, Dupes, and Double Agents

And then there is Obama's Attorney General, Eric Holder, whose job is to enforce all federal laws—including tax laws. He too has a problem paying the taxes he owes. When he was caught failing to pay property taxes on his childhood home in the New York City borough of Queens, which house he and his brother had inherited when his mother died, he said he "wasn't aware" that he had to pay taxes. He owed more than $4,146 in property taxes.

When Holder was caught trying to duck out on his tax bill, his Justice Department spokesman blamed the Attorney General's dead mother. Apparently Holder missed the tax payments "in the last months of [his] mother's life, when she was battling illness." In fact, it took him almost a year after his mother's death to pay the borough of Queens the property taxes he owed—and he only paid up after he was outed in the *New York Post*.[16]

It's hard to believe that Holder simply forgot about his boyhood home on 101st Street in the Queens neighborhood of East Elmhurst. Holder often talked in public speeches and media interviews about the home where he grew up, saying that the neighborhood was a hotspot for black entertainers and personalities. "Jazz legends Louis Armstrong and Dizzy Gillespie lived there, and civil rights leader Malcolm X was a neighbor," the *New York Post* reported.[17] He even bragged about his famous neighborhood in his Senate confirmation hearings for the Attorney General post. But when the taxes came due, he went silent and blamed everyone but himself.

I'll have more on Holder's treasonous tenure as attorney general in chapter 10, "Tyranny of the Anti-Justice Department."

Treasury Secretary Timothy Geithner is another of the Obama tax cheats. He owed a small fortune in back taxes and only agreed to pay up after he was caught by investigators for the U.S. Senate. He realized that he would never be confirmed as Treasury Secretary if he didn't pay his debts to Uncle Sam.

The question became, "Was Obama concerned about putting a tax dodger in charge of the U.S. Treasury, which prints America's money and supervises the Internal Revenue Service itself?"

The answer came out fairly quickly: Not really. Obama's press secretary, Robert Gibbs, dismissed the tax problems as "minor" and said that Geithner's "service should not be tarnished by honest mistakes, which, upon learning of them, he quickly addressed." [18]

Here are the "honest mistakes" the Treasury Secretary—the guy who's in charge of handling the nation's money—made. Geithner failed to pay taxes in 2003 and 2004 while he worked at the International Monetary Fund. The IRS caught up with him in 2006, finding that he owed $17,230 in back taxes and interest. Later, Senate investigators discovered that Geithner also failed to pay taxes for 2001 and 2002, in the total amount of $25,970.[19]

Geithner had all types of excuses. He said that he was entitled to deduct his child's time at overnight camps in 2001, 2004, and 2005 as child care expenses. The IRS wasn't amused. He also tried to deduct the cost of a housekeeper for the three and a half months after her right to legally work in the U.S. had expired.[20] In other words, he was trying to deduct the costs of illegal labor. Finally he claimed there were problems with the software he used to calculate and file his taxes. The software in question was TurboTax, a popular program used by tens of millions of Americans every year without any mysterious "software problems."[21]

Leon Panetta may share more than just a first name with Leon Trotsky, the infamous communist who helped form the Soviet Union in after the bloody revolution of 1917. Disturbing new evidence has surfaced that Panetta, Obama's first CIA director and now Secretary of Defense, may have had a decades-long relationship with Hugh DeLacy, a member of the executive committee of the Communist Party USA with known connections to Soviet agents, "including Victor Perlo of the infamous Perlo spy group, and Frank Coe and Solomon Adler of the equally infamous Silvermaster spy group. DeLacy is also associated with suspected Soviet agent John Stewart Service, of the 'Amerasia' spy case."[22]

Trevor Loudon went through the archives of the University of Washington and discovered evidence of a "cordial, long-term relationship in the 1970s and 1980s" between Panetta and DeLacy. At the time, Panetta was a Democratic congressman from California, known for his liberal voting record, and DeLacy was notorious for his Communist Party USA membership and his fierce anti-Americanism. Loudon discovered that "DeLacy was in regular contact with Leon Panetta, grilling him and regularly asking him for military, defense and foreign policy related information, which Panetta heavily supplied him."[23]

Panetta was also tied to the Institute for Policy Studies, a well-known pro-Soviet think tank based in Washington, D.C. The Institute for Policy Studies vigorously opposed President Ronald Reagan's plans to topple the Soviet-backed Sandinista junta in Nicaragua. Panetta strongly agreed with the institute and even pushed in Congress for a "normalization" of diplomatic relations with communist Nicaragua. Indeed, Panetta as a congressman had a disturbingly pro-Soviet record. He pushed measures to make the USSR and its Eastern European satellite countries into most-favored nations in trade agreements. He voted in favor of turning over the Panama Canal to the government of Panama, which was then an ally of the Soviet Union. He even opposed America's defense pact with Taiwan on the ground that Taiwan acted as a check against communist China.[24]

And what did Panetta think about DeLacy, who met with a string of notorious Soviet and Chinese communist spies? He stood up at DeLacy's 1993 funeral and gave a long speech praising DeLacy. He even saw to it that favorable remarks about DeLacy were placed into the *Congressional Record*.

All of these facts would have been turned up in Panetta's White House background check—and can easily be found by anyone with an Internet connection. We can only conclude that Obama *knowingly and deliberately* put someone who cavorted with communist spies for decades, and who took a hard-line pro-Soviet foreign policy stance throughout his long tenure in Congress, in charge of the nation's secrets and the CIA and the nation's security at the Defense Department. It simply cannot be an accident.

Transportation secretary Ray LaHood may be a former Republican congressman from Illinois, but he is a bought-and-paid-for member of the Chicago political machine. He seems to have two main tasks as Obama's transportation secretary: to get a third runway built at Chicago's O'Hare International Airport at federal taxpayer expense and to oversee the TSA, which gropes grandmothers, inspects infant diapers, and eagerly scans the bodies of attractive young women for hints that they might be terrorists conspiring to blow up commercial airliners. While the TSA is excellent at detecting toenail clippers and bottled spring water in your carry-on luggage, it hasn't been able to catch any terrorists on domestic flights since it was founded. Instead, it is a government jobs program that employs more than 44,000 people, whose union dues to the American Federation of Government Employees go to the reelection efforts of President Obama.[25]

LaHood's biggest accomplishment has been to rob millions of Americans of their time, privacy, and personal space, while rewarding Obama with millions in campaign cash.

While Obama has appointed a cabinet of tax cheats, dupes, and double agents, he has a hard time managing them. Obama asked White House Chief of Staff William M. Daley "to repair badly frayed relations between the White House and the Cabinet." Cabinet secretaries who have important constitutional responsibilities to manage wide-ranging portfolios, from defense and foreign affairs to energy and the environment, complained that the White House and the crowd of czars that Obama appointed have kept them in the dark. "The White House loops people out. The czars keep people from getting in," a senior Democratic official told the *Washington Post*. That official said he has received complaints from at least three agency heads and that the "level of frustration is pretty high."[26]

Cabinet secretaries complained privately to the *Post* that the White House was a "fortress" that micromanaged their agencies and refused to consider practical objections raised by their bureaucracies while "senior administration advisors rolled their eyes at staff meetings at the mention of certain cabinet members."

The solution? To create a czar to manage the cabinet! Obama created a new position of Cabinet Communications Director to "better coordinate with and utilize members of the cabinet."[27]

Usually presidents manage their own cabinets or ask their chiefs of staff to do it. When Obama sees a problem, the answer is to create another level of bureaucracy.

Maybe he doesn't have the time or the patience. President Obama, like many tyrants before him, prefers the trappings of state rather than being trapped into actually making hard decisions. Like spoiled kings who prefer sycophants and fantasies to real governance, Obama prefers his agreeable czars and golf outings to doing the business of state. While Obama meets with his czars weekly and plays golf nearly that often, he can't seem to find time to meet with his cabinet. Obama's history of avoiding any real work and his dislike of hearing views that don't support his own have created a big problem.

Obama plainly doesn't like Secretary of the Interior Ken Salazar. Shortly after the 2010 BP oil spill in the Gulf of Mexico created the largest environmental crisis in a generation, Obama told Salazar that he couldn't talk to the press any longer. Salazar's Interior Department manages tens of thousands of acres of fragile saltwater swamps and marshlands on the Gulf Coast and those protected federal lands were directly affected by the oil spill. Instead, Obama insisted that environment czar Carol Browner would talk to the press.[28] As I've explained to you earlier, Browner is an environmental zealot serving in an unconstitutional position with no direct responsibility for the Gulf Coast ecosystems ravaged by the spill.

Obama also didn't get along with his first Commerce Secretary, Gary Locke, who he later demoted to Ambassador to China. When Obama held meetings with CEOs and other top executives in December 2010, he didn't bother to invite Locke. Nor was the Commerce Secretary invited to attend Obama's speech to the U.S. Chamber of Commerce in February 2011. As a result, one senior official commented that Locke is "not a player."[29] Locke became just another cabinet secretary sidelined by Obama.

Is it any surprise that Obama has not met with many of his cabinet secretaries in more than two years? He does see his czars on a regular basis, and he always has time for union bosses. As I explained in chapter 5, in Obama's first year in office SEIU boss Andy Stern was the single most frequent visitor to the White House. After Stern left the union, AFL-CIO head Richard Trumka visited the White House nearly every day and met with the president on a weekly basis.[30]

Or consider this little item that I dug up. The union-approved head of General Motors, Dan Akerson, knows how to please Obama. The head of America's largest carmaker—which is now 26-percent owned by the federal government thanks to Obama's nationalization of it in 2009—is actually calling for higher gas taxes. Here's what Akerson said: "You know what I would rather have them do—this will make my Republican friends puke—as gas is going to go down here now, we ought to just slap a fifty cent or dollar tax on a gallon of gas. People will start buying more Cruzes and they will start buying less Suburbans."[31]

Akerson says that higher gas taxes would force Americans into smaller, more fuel-efficient cars and might help boost GM's money-losing electric car business. Akerson doesn't have much sympathy for the average motorist. His proposal to hike gas taxes by as much as a dollar a gallon came at a time when gas prices were north of $4 per gallon in the Detroit region. As for people with large families or large toolboxes to transport, too bad.

What about Obama's large ownership stake in GM? "I have nothing but good things to say about them," Akerson told the *Detroit News*.[32]

A Modern-Day Dreyfus Affair

Captain Alfred Dreyfus was a dashing young artillery officer in the French army in 1894, when he was accused of being a spy and handing over French military secrets to their German enemies. The French officer protested his innocence, but his Jewish origins made prosecutors and the public dismiss his pleas. In shackles and leg irons, he was shipped to the tropical penal colony known as Devil's Island off the coast of France's South American colony of French Guyana, and left to languish in solitary confinement. Devil's Island was notorious for its brutal guards, harsh conditions, and malarial mosquitoes. Life expectancy for prisoners held there was barely four years. Most died from a combination of starvation, overwork, beatings, and the tormenting fevers of malaria.[33]

But the Dreyfus Affair, as it was known, was a setup and the young captain was actually innocent. When the real spy was caught, senior French military officers hid the evidence and a military court acquitted the guilty man. The French army brought a second round of charges against Dreyfus based on fake documents created by a crooked French intelligence officer.

By 1898, France's most popular writer, Emile Zola, smelled a rat. He wrote one of the most famous articles in journalistic history, "J'accuse!" In it, Zola demanded that the government reexamine the evidence that had been used to wrongly convict Dreyfus. That same year, Dreyfus was brought back to Paris for the biggest and most controversial trial in French political history. The case dragged on for 12 years, but Dreyfus was eventually found innocent and restored to a position of honor in the French military. He fought bravely in World War I and rose to the rank of lieutenant colonel.[34]

So why am I telling you this bit of ancient history? Simple. As I have been saying on my show from the very beginning of the Dominique Strauss-Kahn case, President Obama and underlings have created a modern-day Dreyfus case against Strauss-Kahn.

Dominique Strauss-Kahn was the head of the International Monetary Fund in May, 2011, when he met a housekeeper at the Sofitel New York hotel in midtown Manhattan. Hours later, he was taken off the plane of a Paris-bound flight at John F. Kennedy International Airport by New York City police officers and made to walk in handcuffs before the television cameras of the world. To his shock, he was accused of raping the hotel maid, an immigrant from the West African country of Guinea. When he was arraigned before a New York judge, he pleaded not guilty. Bail was set at $1 million and he was put under house arrest in a friend's condo in Greenwich Village.

Soon he had to resign as head of the IMF and spend tens of thousands of dollars to defend himself against criminal charges that could put him in a federal lockup for the rest of his life.

So far Obama's plan seemed to be working. With Strauss-Kahn out of the way, the new head of the IMF was Christine Lagarde, a Chicago lawyer who had given the maximum legally allowed amount to the 2008 Obama presidential campaign.

But by July, 2011, the case was falling apart, just as I predicted. And just like the case of Captain Dreyfus.

I had been predicting this since the moment Strauss-Kahn was arrested on May 14, 2011. Here's what I predicted on my radio show on May 16: That the maid was down-and-out and that after Strauss-Kahn solicited her for sex, she consented. But then the maid cried "rape" in order to make fame and fortune. I repeatedly reminded listeners that in America, people are innocent until proven guilty. I also reminded listeners that as the front-runner for the presidency of France, Strauss-Kahn has political enemies including the current president of France, Nicolas Sarkozy, who wants to keep his job.

Since Sarkozy knew Strauss-Kahn's well-known weaknesses for women, he knew perfectly well how to set him up. Then I played a clip from *The Godfather Part II* in which mafia boss Michael Corleone sets up a U.S. senator by putting a dead prostitute in his hotel room. Now the senator can be blackmailed into doing what the family wants. This is exactly what the Strauss-Kahn setup smells like, I said.

If you were a rich and famous guy staying at a New York hotel, I said, a girl could have been "delivered with his morning orange juice." There is no reason for Strauss-Kahn to risk his freedom, his fortune, his reputation, and his future by raping a chunky hotel maid from an impoverished West

African backwater. He could have had a sleek model sent up just like former New York governor Eliot Spitzer did.

On July 1, again I was right. Here's what I said on my radio show:

> Dominique Strauss-Kahn, former head of the IMF, was released today on his own recognizance. No doubt the case is going to fall apart. When you read the details, it becomes clear that I was right that this was a setup from the get-go. So who set him up? He was the head of the IMF, and was about to become the president of France, running against Sarkozy, who was falling in the polls. I speculated, as did many others with a brain, that Strauss-Kahn was set up in the hotel room. No doubt this was consensual sex and then she screamed rape. They knew his proclivities and that Strauss-Kahn could not control himself.

So who did it? Well, you can guess who did it, you could say Sarkozy's people did it, but you will never prove it. You could say that this woman Christine Lagarde was involved. She is the new head of the IMF and a protégé of Sarkozy. In fact, if you investigate Christine Lagarde, there are many interesting details in her past. For example, there's a scandal in France where she arranged a questionable 285 million euro payout to a French business-man, Bernard Tapie, in 2007.

Also, she worked for William Cohen. Remember him? Senator from Maine. Later Defense Secretary under Bill Clinton. Also well-known as a good friend of Timothy Geithner. Lagarde also rose to the head of a top Chicago law firm. Very interesting for a woman born in the United States, but from France, really. How did she rise to the top of a Chicago law firm? Let's use Chicago now as a starting point. Chicago: Rahm Emanuel, Bar-rack Obama, and now the head of the IMF? It seems to me that Chicago is running the world and the world economy. Like I say, this is bigger than the Dreyfus Affair and it is worthy of a movie.[35]

And I have been repeatedly vindicated by events. Here are some details you haven't heard from the mainstream media. The hotel maid has admit-ted to lying and fabricating evidence on her application for asylum to the United States in 2004, making her essentially an illegal immigrant.[36] Next the hotel maid phoned her boyfriend, who was held in a federal detention center pending a long list of charges, including possession of 400 pounds of marijuana.[37] The *New York Times* made it seem like the guy was just being held for "immigration violations."

Next, I found out that the maid's phone call was in a "unique dialect of Fulani," a tongue of her native West African country of Guinea, in the hopes that the prison guards would not be able to translate her words. But all phone calls to federal prisoners are recorded and the New York district attorney was motivated enough to hire a special translator. A few days later, the translator brought back shocking news. The maid had told her jailed boyfriend, "Don't worry, this guy has a lot of money. I know what I am doing."[38]

Then it emerged that she had lied about being raped before. She said that she had been gang-raped in Guinea and that is why she deserved asylum in the United States. That story turned out to be false and cribbed from another West African immigrant who had used the lurid tale to win asylum in the United States. So she decided to go with the lie that worked before, and witnesses told prosecutors; she would often use a tape recorder to play back the details of her fake gang rape so that she could tell the false tale in her sleep.

While prosecutors initially believed the maid's story and sheltered her under a fake name in a Brooklyn hotel, her activities in that hotel raised disturbing questions. She entertained a string of men, ranging from wealthy businessmen she had met at the Sofitel to "counterfeit merchandise hawkers and livery cab drivers," the *New York Post* reported. While some of her clients paid her in cash, others gave her expensive jewelry, which is easier to hide from the prying eyes of the IRS. "While she was under our supervision, there were multiple 'dates' and encounters at the hotel on the DA's dime," a source told the *New York Post*. "That's a great deal for her. She doesn't have to cover her expenses."[39]

When prosecutors would ask her hard questions about details that didn't check out, she would spontaneously start crying and even roll on the floor. And sometimes she would disappear for days, refusing to answer her phone when prosecutors called and failing to answer the door when they knocked.[40]

Next, it turned out that she had lied on her federal and state income tax returns and had unexplained cash deposits of $100,000 into her bank account while she worked as a hotel maid under a union contract that, even with overtime, does not pay $100,000 a year.[41] The New York district attorney now suspects that that nest egg was generated by "proceeds from sex for money exploits."[42] The *New York Times* concluded: "Little by little, her credibility as a witness crumbled—she had lied about her immigration, about being gang raped in Guinea, about her experiences in her homeland and about her finances, according to two law enforcement officials. She had been linked to people suspected of crimes. She changed her account of what she did immediately after the encounter with Mr. Strauss-Kahn."[43]

While the case against Strauss-Kahn was falling apart in New York, the damage was still being felt in Paris. He was no longer head of the IMF and his ability to win an election in France was in doubt. *Le Parisienne*, a Paris-based daily newspaper, surveyed French voters about the possibility of Strauss-Kahn returning to electoral politics. Strauss-Kahn's support among Socialist Party members (his political party) has fallen from near unanimous support to 55 percent. Among French voters of all political parties, only 49 percent of voters believed that Strauss-Kahn should run for office, while 45 percent did not. Before the hotel incident, polling for Strauss-Kahn was so strong that he was considered to be a shoo-in for the French presidency.[44]

Days after the U.S. case started to publicly fall apart, a young Frenchwoman, Tristane Banon, announced she would be filing a lawsuit alleging attempted rape ten years ago during a 2002 book interview with Strauss-Kahn.[45] You want me to believe this woman waited a full decade to seek justice, but couldn't delay more than a few days after the collapse of the New York rape case to announce her intention to press charges? Strauss-Kahn can look forward to an endless string of false accusations as long as he remains a potential rival to Sarkozy.

And let's not forget about Christine Lagarde. As I said, I dug into her weird background, too. Remember, she is the new Managing Director of the IMF. She is the first woman ever to head the IMF and has held a number of cabinet-rank posts in the French government, including Minister of Economic Affairs, Finances and Industry, Minister of Agriculture and Fishing, and Minister of Trade. While she was born in Paris into a family of left-wing academics, she went to a pricey prep school in Bethesda, Maryland, the Holton-Arms School. While in that elite high school she worked as an intern for then-congressman William Cohen, who was later handpicked by President Clinton to be Secretary of Defense.

After college and graduate school in France, she returned to the United States to work at Baker & McKenzie, a global law firm based in Chicago. She made partner in 1987, joined the law firm's executive committee in 1995, and became the law firm's first female chairman in 1999.[46] She is a yoga-practicing vegetarian who never drinks alcohol, according to Paris Match.com. The divorced mother of two "is said to enjoy being depicted as a dominatrix who whips bankers."[47]

Tax cheat Timothy Geithner has nothing but good things to say about Lagarde, such as that "her lightning quick wit, genuine warmth and ability to bridge divides while remaining fiercely loyal to French interests, have been a source of admiration."[48]

So, at a time when the falling dollar and rising oil prices mean that Obama badly needs a friendly voice at the head of the IMF, he's got one. Strauss-Kahn, as a guardian of the IMF's finances, was reluctant to sanction Obama's wild printing of money. Under Obama, the Federal Reserve Bank went from holdings of $800 billion in 2009 to more than $2 trillion in 2011. That's why central bankers and large investors around the world have been dropping dollar-denominated debts and exchanging dollars for more secure currencies—they believe that the U.S. dollar is about to collapse.

The job of the IMF is to maintain the stability of major global currencies, especially the U.S. dollar. Strauss-Kahn did not like what Obama was doing with the American money supply and repeatedly said so. Now he is gone. In his place is a friendly French-born Chicago lawyer who, we cannot forget, donated the maximum amount allowed to Obama's 2008 political campaign and who has ties to the inner circle of Obama's Democratic Party, including Treasury Secretary Geithner, and to former Defense Secretary William Cohen. Lagarde, so far, has been pointedly silent about Obama's dangerous increase in the U.S. money supply. An early test of Lagarde's loyalty to Obama will be the president's demand for millions of dollars to be sent to the Egyptian government, which is jointly run by an Egyptian army junta and elements of the radical Muslim Brotherhood.

Kickbacks for Obama's Cronies

When he announced his candidacy for president at the birthplace of Abraham Lincoln in February 2007, Obama said he was strongly opposed to "cynics, lobbyists, and special interests." He couldn't have sounded more high-minded. "They write the checks and you get stuck with the bill. They get the access, while you get to write a letter. They think they own this government. But we're here today to take it back. The time for that kind of politics is over. It is through. It's time to turn the page right here and right now."

Once in power, Obama began to reward those very same cynics, lobbyists, and special interests who wrote him checks. More than 200 of Obama's biggest campaign contributors have won places of power inside the Obama administration or had their companies rewarded with federal contracts, according to a report by *iWatch News*.[49] *iWatch* is a project of the Center for Public Integrity, a nonprofit group that does independent investigative reporting.

Take the case of Donald H. Gips. I investigated him and found a rat. He is the Vice President of Level 3 Communications LLC, a Colorado-based telephone service provider. He raised more than $500,000 for Obama in 2008, and two other executives at his firm kicked in a total of $150,000.

Gips's reward?

He was put in charge of selecting people to work in the Obama White House and later Obama made him Ambassador to South Africa. As for Level 3 Communications, it received $13.8 million in stimulus money. That's a return 28 times greater than the firm's total $650,000 investment. You can't do that in the stock market, even if you are George Soros. But you can with Obama. It's called Obamanomics.

That's how politics in Chicago works. Just ask former Illinois Governor Rod Blagojevich, who was caught trying to sell Obama's vacated Senate seat to the highest bidder. Wiretaps show how he played hardball with negotiations:

> If . . . they're not going to offer anything of any value, then I might just take it. . . . Unless I get something real good for [the Senate

candidate], . . . I'll just send myself, you know what I'm saying. . . .
I'm going to keep this Senate option for me a real possibility, you
know, and therefore I can drive a hard bargain. . . . [The Senate seat]
is a [expletive] valuable thing, you just don't give it away for nothing. . . . I've got this thing and it's [expletive] golden, and, uh, uh,
I'm just not giving it up for [expletive] nothing. I'm not gonna do it.
And, and I can always use it. I can parachute me there.[50]

Gips is not an exception. Almost 80 percent of those who raised more
than $500,000 for Obama were given "key administration posts, as defined
by the White House."[51]

The other major donors, who didn't want to take a pay cut for work-
ing for the federal government, were rewarded with private meetings in the
White House and invitations to its glitzy dinners. More than 3,000 White
House visits have been linked to Obama campaign donors, according to
White House records.

Of course, President George W. Bush also gave jobs to donors, mostly
making them ambassadors to South American and Pacific island nations.
Overall, Bush gave 200 donors positions in the federal government. But
Bush did so over eight years, while Obama rewarded more than 200 donors
in just two years, according to Public Citizen, a left-wing watchdog group
founded by Ralph Nader.[52]

It's the best administration money can buy.

And selling sensitive government positions to Democratic donors is
about to increase sharply. An Obama administration memo leaked to the
Huffington Post reveals how they are going to handle a donor named Ed
Haddock, who raised more than $200,000 for Obama in 2008. Haddock is
the CEO of Full Sail University, a for-profit school in Orlando, Florida, that
is under attack from Obama regulators. The memo notes that Haddock
stopped being "helpful" in 2009 and may give to former Massachusetts
Governor Mitt Romney, Republican candidate for president. The memo
was written by Jessica Clark, Florida Finance Director for Obama's 2012
campaign, to top White House aides. The memo advised how the White
House should treat Ed Haddock: "You should engage Ed on his concerns
and tell him you want an ongoing relationship that seeks to hear his ideas
and concerns, even if in the end we don't always agree."[53]

The memo provides further evidence that Obama treats the White House as a cash register for political fund-raising. The memo itself may also be illegal. Under the Hatch Act, federal employees, including those who work at the White House, are banned from political fund-raising on the taxpayer's dime and barred from engaging in partisan politics.[54]

And have you heard about Siga?

The company represents another example of how Obama uses stimulus money as his own slush fund to pay back donors. I'll tell you about a number of Obama administration scandals in coming chapters, including the LightSquared scandal in chapter 7, "Tyranny of the Egghead Wars," and the administration's green energy scandals in chapter 9, "Tyranny of Green Energy."

But the Siga scandal may be the poster child for this administration's corrupt crony socialism.

Nicole Lurie is a former deputy assistant at the Department of Health and Human Services. Lately she's been employed as an Obama appointee heading "biodefense planning" at HHS. What has she been investigating? Smallpox. That's right, she's been looking into the possibility that there might be a smallpox epidemic in the near future if terrorists unleash the vaccine as part of a biological warfare campaign.

You might recall that smallpox was eradicated around the world in 1978.

So Siga Technologies Inc., headed by billionaire and longtime Democratic Party donor Ronald O. Perelman, went ahead and developed a smallpox vaccine, then used its lobbying muscle to grab a no-bid contract for 1.7 million doses of a drug that prevents smallpox. In other words, a drug that has no practical use whatsoever. It also happens to be a drug that has never been tested on humans. And its estimated shelf life is a little over three years.

Perelman has a lot of clout in governmental circles. Siga complained that the contract bid was being held up by HHS officials, so those officials were replaced. In May 2011, Siga was the only company to submit a proposal to deliver the vaccine. Nobody else was stupid enough to work on developing a drug for a disease that no longer exists. They didn't realize what they were missing.

Lurie was central in the negotiations, coordinating contacts between White House officials, HHS, and Siga. She also had a hand, allegedly, in making sure HHS replaced officials who were questioning the need for the contract with others who were open to Siga's play.

In the end, HHS agreed to pay $433 million to Siga. That amounts to $255 for each of the 1.7 million doses, doses that cost $3 each to produce. Lurie explained that the contract was awarded based on the estimated value of the benefit that might be realized from the drug, not the cost to produce it.[55]

Nearly every time you turn over a rock in this most scandal-ridden of presidential administrations you find either an egghead trying to implement a radical agenda, an incompetent who is screwing up his vital duties, or an anti-American playing for the other team. And we haven't even gotten to Obama's Justice Department, headed by Attorney General Eric Holder. That is such a deep mine of outrageous actions and unqualified dupes that it will take another chapter to begin to tell the story.

CHAPTER 7

Tyranny of the Egghead Wars

When I was a kid in the 1950s and early '60s, I would often help my father out in his antiques store on the Lower East Side of Manhattan. It was a long ride in from Queens, but I went along with my father because it was like a passport to a different world, where I saw all types of strange characters. Everyone, rich or poor, wore a suit and a hat. I met lawyers and doctors, artists and sailors, stock traders and European art collectors. Once, I even met someone from the freak show at the Ringling Brothers circus.

But, as far as I know, I never met an Ivy League professor.

Of course, I heard about them. Respectfully. It's hard to believe now, but back then, we looked up to them as great scholars, wise men.

This was before the Ivy League professors took over the Lyndon Baines Johnson administration and launched Medicare, which all of the sheeple thought would help the poor but instead is now bankrupting the government and the middle class.

It was before the Ivy League professors put so many restrictions on our troops ("proportional response," they called it then) that we lost the Vietnam War and under the Obama administration are now in the process of losing the war against Islamic terrorists in the Middle East.

It was before the Ivy League theorists wrecked the economy in the 1970s, causing the falling dollar, rising national debt, skyrocketing personal bankruptcies, and the worst unemployment rates since the Great Depression. At the same time, the auto and airline companies were going broke and demanding government handouts, while climbing gas and food prices ate up your paycheck—if you were lucky enough to have a job.

Sounds familiar, doesn't it?

It took Ronald Reagan, a graduate of Eureka College in the tiny town of Eureka, Illinois, to fix the mess the Ivy Leaguers had made of the 1970s.

Since Reagan, it hasn't much mattered which party was in power. Ivy Leaguers have had a plan for everything.

When they arrive at the White House, watch out.

Ivy Leaguer George H. W. Bush (Yale) brought us the savings-and-loan meltdown, which cost taxpayers over $100 billion.

Bill Clinton (Yale Law School) gave us Monica Lewinsky, attempted to pass socialized medicine in the form of Hillarycare—the precursor to Obamacare—and missed at least eight chances to take out Osama bin Laden before 9/11.

From George W. Bush (Yale, Harvard Business School) we got the Iraq War—which he and his advisors almost managed to lose—and the economic crash of 2008.

But the Ivy League superstar who puts the others to shame is Columbia University/Harvard Law School graduate Barack Hussein Obama.

Obama brought us a recommitment to the war in Afghanistan, where U.S. casualties have risen to record highs in the past year. He's pulling the plug on our fighting forces in Iraq, very likely making sure that after our commitment of thousands of U.S. servicemen's lives and billions of dollars, Iran will regress to become another in the growing number of Islamist dictatorships that are emerging as part of the Arab Spring that Obama has encouraged.

Along the way, he committed the United States to war in Libya and to sending American troops into the central African country of Uganda. I'll explain his reasons for doing this later in the chapter.

Because neither Obama nor the fellow Ivy League eggheads he's brought into his administration in record numbers have ever served in the military or know the first thing about what's involved in deploying troops or making decisions in the heat of conflict, Obama dithered for weeks after the uprising had begun in Libya. During that time, Ghadafi sent tanks and machine gun-wielding soldiers against unarmed demonstrators in Tripoli. Soldiers followed fleeing students into stores and shot them in the head as they pleaded for mercy.

Still, Obama refused to even make a public statement.

The president was away on his midterm spring break in Brazil with his family when his administration finally came to a decision. That was shortly after he'd filled out his brackets for the 2011 NCAA men's basketball tournament on ESPN[1] and just before he promised the Brazilian oil company Petrobras $2 billion to drill for oil off their coast and that we'd be their "best customer" for the oil that was produced.[2]

He briefly interrupted his Brazil trip to change course from noninvolvement, announcing that we would intervene in Libya in order to help the insurgents after all. The key to the U.S. intervention in Libya can be found in the president's words when he revealed the change in plans: We cannot stand "idly by while a tyrant tells his people there will be no mercy."[3]

Of course, that's precisely what Obama had been doing for weeks.

Don't think that the president suddenly grew a spine, though.

He wasn't involved in the decision at all.

All the President's Women

In the ivory tower occupied by the academics in his administration, real men are not welcome, so it fell to three women—Secretary of State Hillary Clinton, Ambassador to the United Nations Susan Rice, and leftist policy wonk Samantha Power—to actually decide that we should go to war against yet another country in the Middle East.[4] The decision process reveals just how flimsy American foreign policy has become in the testosterone-free zone known as the Obama White House.

In overriding the president and changing U.S. policy so that we could justify intervening in Libya's internal affairs, Power, Rice, and Clinton based their rationale not on our nation's vital security interests but on an obscure doctrine—championed by international leftist financier and political agitator George Soros—called "responsibility to protect." It's the way eggheads conduct military policy, and it shows how misguided America's military priorities have become in the Age of Obama.

In this case, their rationale was this: It was possible that civilians in Libya might be at risk for being killed by Moammar Ghadafi's military forces when Ghadafi responded to the rebels who threatened his regime, so the United States needed to intervene to make sure that didn't happen.

That's not the way I see it.

Let me make it clear what the "responsibility to protect" doctrine really means.

Here it is in plain English: America's national interests aren't about protecting ourselves or our allies, keeping the sea lanes open and the oil flowing for our economy, or even stopping dangerous radicals like the Iranian mullahs from getting the atomic bomb. What matters is preventing the deaths of people with whom we have no tie and no treaty.

If we get any possible benefit from a military intervention, then the policy must be selfish and wrong. On the other hand, if it costs hundreds of millions of dollars and risks American lives for no long-term gain, then it is a "responsibility to protect" and we must do it.

I told you that these Ivy Leaguers have everything backwards.

In reality, the doctrine is a convenient excuse to selectively intervene in the internal affairs of our Middle Eastern allies in order to hasten their downfall and hasten the formation of Islamist governments.

In other words, Obama doesn't see our "allies" like I do.

As I'll explain in the next chapter, for Obama our old friends are now our enemies and our former enemies are our new friends.

Getting back to the "responsibility to protect" and how it came to be the principle on which we entered another war: Samantha Power was a strong proponent of this misguided doctrine, and it was she who convinced Hillary Clinton to change her position on entering the combat in Libya. Like the president himself, the three women who declared war on Libya are all products of the leftist academic incubator that passes for an educational system in the United States.

The first thing you need to know about Power is that she's married to Cass Sunstein. I explained to you in chapter 6 about Sunstein. He's the wacko leftist Harvard Law School graduate and former University of Chicago professor who was appointed by Obama to the position of regulatory czar. He hates gun owners and the Second Amendment, which protects the right of individuals to own firearms. He also opposes hunting, a legal pastime of some 12 million Americans. He's described the suffering of animals inflicted by hunters as "morally equivalent to slavery and the mass extermination of human beings."

In a public speech in 2007, he dismissed the Second Amendment as a sign of the paranoia of the Founding Fathers against standing armies. He seems to imply that because the United States has a standing army and a national guard, it doesn't need to have guns in the hands of ordinary citizens. He talks as if the meaning of the Second Amendment, which guarantees the rights of citizens to "keep and bear arms"—or in modern words, "own or carry" guns—has been twisted by the National Rifle Association and other civil rights groups. He argues that since the United States has a standing army of its own, the Second Amendment is meaningless.[5]

Samantha Power shares Sunstein's and Obama's leftist credentials and their political views. She also graduated from the bastion of leftists that Harvard Law School has become. Power was part of the Obama presidential campaign until she called Hillary Clinton "a monster" and had to resign. After the election, she took a position as an Obama foreign policy advisor.[6]

She has been a longtime proponent of Soros's "responsibility to protect" doctrine. In fact, her position on the issue won her a Pulitzer Prize. Her prize-winning book, *A Problem from Hell*, is an academic's view of genocide. It was financed in part through a grant from Soros's Open Society Institute. Power, as she explains in the acknowledgments section of her book, interviewed "hundreds of men and women" who had survived genocide. It is from this perspective that Power became a strong advocate of the "responsibility to protect" justification for intervening in the internal affairs of a nation despite the fact that there is no vital interest for the United States.

Secretary of State Hillary Clinton, another Ivy League law school egghead—she went to Yale—is the second member of this unholy trio. During her husband's presidency, the Clintons requested that the subject of Hillary's senior thesis not be made public. That's because the thesis—titled "THERE IS ONLY THE FIGHT—An Analysis of the Alinsky Model"—

was highly favorable to the methods of radical leftist political organizer Saul Alinsky, who Hillary described as "a man of exceptional charm."[7]

As I explained in chapter 5, Alinsky is known as the father of community organizing, and we have him to thank for our community organizer president's views on many issues. Both Clinton and our president are Alinsky disciples, and they've never forgotten the lessons they learned from the master.

The influence of the radical left doesn't stop there.

Susan Rice, our ambassador to the United Nations, was mentored by Bill Clinton's Secretary of State Madeleine Albright. When Albright was recently confronted in an interview by Lesley Stahl about the fact that the U.S. military did not intervene to save the lives of half a million children who died at the hands of Iraqi dictator Saddam Hussein, she replied, "I think this is a very hard choice, but the price—we think the price is worth it."[8]

In other words, the "responsibility to protect" applies to some victims but not all?

Confusing, isn't it?

I've figured out why Hillary changed her tune, too. She remembered how much her husband was pilloried for leaving the poor Rwandans to die in a genocide. A single radio tower in Rwanda broadcast the locations of the people marked for death. If Clinton had bombed that tower, thousands would have been saved. He did nothing and has been criticized ever since. So she was not going to make that mistake again. She immediately began using the magic phrase a "responsibility to protect."

Once Hillary Clinton and Susan Rice joined Samantha Power in the fight to intervene in Libya, Obama was helpless to object.

He's no match for three women.

It wasn't until nine days after he had unconstitutionally committed us to a third Middle Eastern war that Obama bothered to address the nation as he made his own "mission accomplished" speech. He made sure to schedule the speech for 7:30 in the evening, because ABC told the White House it wouldn't carry the speech if it conflicted with *Dancing with the Stars*. At least ABC has its priorities right.

In the speech, he insisted that "when our interests and values are at stake, we have a responsibility to act."[9]

As I've told you repeatedly, in reality, we had no national interest in the Libyan conflict.

By "responsibility to act" Obama meant "responsibility to protect."

It's the same "responsibility to protect" doctrine that was nowhere to be found when Mahmoud Ahmadinejad was murdering civilians in the 2009 Iran uprising, yet it magically appeared, forming the basis of the Obama feminists' rationale for going to war in Libya.

The doctrine was dreamed up by two leftist intellectuals, Ramesh Thakur and Gareth Evans. Like everyone else Obama listens to, they're academics with no real-world experience to back up their theories. Thakur was Assistant Secretary-General of the United Nations for nearly a decade, and he's now employed as a professor at the University of Waterloo in Canada. Evans is a politician and academic who currently serves as Chancellor of the Australian National University. He was active for 21 years in Australian politics, serving as a liberal cabinet minister from 1988 to 1996. He's also been active in the United Nations, and recently released a book titled *The Responsibility to Protect: Ending Mass Atrocity Crimes Once and for All.*

Both Thakur and Evans are cronies of Soros. Evans is a former president of Soros's International Crisis Group. The ICG describes itself as "the world's leading independent, non-partisan, source of analysis and advice to governments, and intergovernmental bodies like the United Nations, European Union and World Bank, on the prevention and resolution of deadly conflict." [10] Soros's partner in the ICG is Zbigniew Brzezinski, a committed Israel hater and former National Security Advisor to another Israel hater, former president Jimmy Carter. Brzezinski supervised the Carter administration conspiracy that resulted, in 1979, in the fall of the Shah of Iran and the implementation of the Islamist dictatorship that currently threatens Israel and the stability of the entire Middle East, and that remains an existential threat to the United States itself.

The true purpose of the ICG is exactly the opposite of its stated purpose. It seeks nothing less than the political downfall of regimes in Muslim countries that maintain friendly relations with the United States, with the ultimate purpose of reducing U.S. power and influence and promoting the power and interests of Islamists.

What happened based on Thakur and Evans's theory is criminal. First, the United States had made its disguised declaration of war on Libya without congressional approval. Then it sent CIA agents into Libya and committed U.S. warplanes to shut down Ghadafi's air force.

Although the president had falsely maintained that our "interests" were at stake, Defense Secretary Robert Gates explained that Libya is not "a vital interest" for the United States.[11] And where Obama declared that going to war in Libya also depends on our "values"—by which I assume he means our "humanitarian values"—he neglected to point out that we routinely turn our backs when tyrants murder their citizens to keep order.

Only one person, New York congressman Charles Rangel, had the courage to speak out about the fact that Obama's ordering U.S. involvement in Libya was unconstitutional.[12] As they have in so many other matters of concern—from taking real measures toward balancing the federal budget to defunding health care—Congress ignored the constitutionality issue, abdicated its sworn duty to uphold the Constitution, and ceded unchecked power to the president.

A month after Obama had initially chosen to ignore the conflict in Libya, and based on a suspect policy promoted by all the president's women that relegates U.S. vital interests to secondary status, the U.S. joined a coalition of NATO nations to establish a no-fly zone over Libya.

About a week after the U.S. had joined the NATO coalition in creating the no-fly zone over Libya, Defense Secretary Gates and Joint Chiefs of Staff Chairman Admiral Mike Mullen announced without warning that the U.S. was pulling its planes out of the operation and handing complete control of the mission over to NATO. The announcement came as Ghadafi's military completed a fourth straight day of forcing the rebels into retreat and regaining much of the territory that had been lost as a result of the no-fly zone going into effect, despite the best NATO efforts.[13]

Gates washed his hands of the matter, saying, "My view would be, if there is going to be that kind of assistance to the opposition, there are plenty of sources for it other than the United States." At the same time, the U.S. sent CIA operatives into Libya to assess the situation and determine the feasibility of further action, including arming Libyan rebels. This apparently didn't count as our having "boots on the ground," something the Obama administration had repeatedly said would not happen.[14]

In the meantime, things became so muddled among the leaderless

NATO forces that they couldn't even decide which side they were on. Since they too were committed to the "responsibility to protect" doctrine, which says to attack whichever group kills civilians, it wouldn't be long before NATO had to declare war on itself. It caused "collateral damage" in the form of dozens of civilian deaths at its own hand when it bombed and strafed Ghadafi's forces as part of establishing a no-fly zone over Libya.

U.S. Army General Carter Ham, who initially led NATO forces in the Libya mission, explained, "I cannot be sure that there have been no civilian casualties. What I can be sure of is that we have been very, very precise and discriminate in our targeting." [15]

"We've been conveying a message to the rebels that we will be compelled to defend civilians, whether pro-Ghadafi or pro-opposition," said a senior Obama administration official. "We are working very hard behind the scenes with the rebels so we don't confront a situation where we face a decision to strike the rebels to defend civilians." [16]

Behind the scenes, while all this was going on, Ghadafi sent a letter to our president, in which he addressed Obama, "Our dear son, Excellency, Baraka Hussein Abu oumama." The letter goes on to say, "We have been hurt more morally [than] physically because of what had happened against us in both deeds and words by you. Despite all this, you will always remain our son whatever happened. We still pray that you continue to be president of the U.S.A. We Endeavour and hope that you will gain victory in the new election campaigne." [17]

Obama left the task of responding to the letter to his Secretary of State, who answered mysteriously, "I don't think there is any mystery about what is expected from Mr. Gadhafi at this time."

Militant black nationalist and Ghadafi/Nation of Islam sympathizer Louis Farrakhan, Obama's friend and a longtime supporter of Ghadafi, shed additional light on the situation and on Obama's loyalties when he said, "I love Moammar Gadhafi, and I love our president. . . . It grieves me to see my brother president set a policy that would remove this man not only from power, but from the earth." [18]

Still can't figure out what the U.S. policy was?

You're not alone.

The Obama administration continually sent mixed messages about our involvement in Libya. Press Secretary Jay Carney expressed it this way: "Our goal of having Gaddafi step down, take himself out of power or be removed from power, is a non-military goal." [19]

The "Responsibility to Protect" Doctrine Helps Islamist Radicals in the Middle East

I've found that liberals love the word *rebel*. To their teenage sensibilities, it conjures up images of 1950s actor James Dean in *Rebel Without a Cause*.

There's something romantic about the word.

Rebels are worth fighting for.

Their use of the word is also one more way they show their ignorance.

The problem is that in Libya, the "rebels" were infiltrated by Islamist radicals, members of al Qaeda and the Muslim Brotherhood who see the conflagrations around the Middle East as prophetic. To Islamist dictators and jihadis, the current uprisings are nothing less than a sign that the coming of the 12th Imam, the Mahdi, is imminent.

We're supposedly fighting a war against these radical Islamic psychos.

While no one in our government is willing to confront this, Islamic radicals are not going away. They have been an integral part of the resistance to the decades-long rule of such dictators as Saddam Hussein in Iraq, Hosni Mubarak in Egypt, Moammar Ghadafi in Libya, and the Saudi royal family, leaders who implemented Western-style military dictatorships or family/tribal repressive monarchies that ran directly counter to the Islamist radical dictatorship in Iran.

As everyone but the leftists in the Obama administration knows, Iran is the center of the Islamist move to destroy Israel and the United States and spread Shariah law around the world. Iran's purpose is nothing less than total domination of the West. The "responsibility to protect" doctrine and the selective way it's being applied by the United States is helping to pave the way for a broad takeover of Middle Eastern countries by radical Islamist forces.

If you don't believe that, take a close look at what's going on behind the scenes.

You remember Mohammed ElBaradei, right? ElBaradei is the guy who led the inspections of Iran's nuclear program when he was the head of the International Atomic Energy Agency (IAEA). In 2007, ElBaradei called for everyone to take a "timeout regarding the Iranian nuclear issue." He hoped that Iran would suspend its nuclear "enrichment activity" and "go immediately to the negotiating table." [20]

As it turned out, ElBaradei was just stalling for time. A year later he gave the Iranian nuclear program a clean bill of health, saying that Iran's explanations of its suspicious nuclear activities "are consistent with [the IAEA's] findings [or at least] not inconsistent." [21] The *Wall Street Journal* described ElBaradei's agenda in blunt terms: "In reality, he is a deeply political figure, animated by antipathy for the West and for Israel on what has increasingly become a single-minded crusade to rescue favored regimes from charges of proliferation." [22]

In the meantime, yet another Iranian facility for the production of centrifuges used in the enrichment of weapons-grade uranium was discovered. [23]

ElBaradei's pro-Islamist, anti-Israel agenda has made him a perfect mouthpiece for George Soros and the spread of the "Soros Doctrine" in the Middle East. In fact, he's on the board of Soros's International Crisis Group, and he was handpicked by Soros to be his representative in Egypt. On January 18, 2011, before the uprising in Egypt had begun, ElBaradei, in what was a veiled call to action for Islamists, warned that a "Tunisia-style explosion" could occur in Egypt.

Obama wasn't standing idly by while all this was happening.

The U.S. knew as early as December 2008 that groups opposed to the Mubarak regime were developing a plan to overthrow the Egyptian government. They received the information from a young dissident who the U.S. had sponsored to attend a meeting for international political activists that took place in New York City. [24]

In addition, according to documents exposed by WikiLeaks, U.S. Ambassador to Egypt Margaret Scobey was aware of the plans of the Mubarak opposition group. Leaked documents also show that while the United States publicly supported the Mubarak government, U.S. Embassy officials continued to communicate with the activist in question throughout 2008 and 2009. [25]

After Mubarak left office, ElBaradei emerged as a key figure in a "shadow parliament" that formed in Egypt.[26] The shadow parliament consisted of opposition leaders who were trying to develop plans for a transition to a new regime through "free and democratic" elections. Included in the group were representatives of the Muslim Brotherhood, the radical terrorist group responsible for the assassination of Egyptian president Anwar Sadat in 1981 and the seed group for other Islamist terrorist organizations, including al Qaeda.[27] The Muslim Brotherhood seeks nothing less than a government based on Islamist principles, including the implementation of Shariah law and waging jihad against the West.

ElBaradei, given the fact that he looked the other way when inspecting Iran's nuclear facilities, is very likely a puppet of the Iranian regime. In April 2009, ElBaradei told the press that "more U.S. engagement with Tehran's leaders would increase regional security."[28]

The list of the people on the board of directors of the ICG along with ElBaradei reads like a Who's Who of international anti-American, anti-Israel zealots. It includes Javier Solana, one of the most powerful figures in the European Union. Because of his Marxist sympathies and his support for the regime of Cuba's Fidel Castro, Solana was at one time on the United States' subversives list. Former Clinton administration National Security Advisor Sandy Berger, who once smuggled incriminating documents out of the National Archives by hiding them in his clothing, is another board member, as is General Wesley Clark, once fired from his NATO command position.

Did you think that the people's uprisings in the Middle East were being staged by Arabs seeking truly democratic governments?

Do you think that the people of Tunisia and Egypt share the same belief in the effectiveness of nonviolent protest that the wide-eyed innocents in the Obama administration hold?

Let me set you straight.

The U.S. closed its eyes as genocide was committed on a massive scale in the Middle East, as long as it wasn't committed by an ally of the United States. So it was no surprise that we took the side of the rebels in Libya and against Ghadafi. The Libyan resistance is heavily infiltrated by radical Islamist jihadists.

And if you need proof of the fact that a substantial number of the "rebels" in Libya are al Qaeda affiliates, look no further than the "Sinjar documents." In 2007, our forces in Iraq seized a computer with biographical records, including the countries of origin, of more than 700 insurgents who came to Iraq to fight against the American "invaders." While more than 40 percent of the recruits came from Saudi Arabia, nearly 20 percent came from Libya.[29] Many of them came back to Libya and joined the fight against Ghadafi.

Within a week of the beginning of NATO and U.S. involvement in the war in Libya, it appeared that things had changed dramatically in the rebels' favor. Euro and U.S. fighter jets had shut down Ghadafi's air force, and the balance of power seemed to have shifted to the insurgents. In other words, by shutting down Ghadafi's airpower, the "allies" made way for such groups as al Qaeda and the Muslim Brotherhood to make advances toward their goal of establishing another radical Islamist government in the Middle East.

Sorobama and the Assassination of Moammar Ghadafi

But maybe the event that best sums up U.S. involvement in the Libyan war is the killing of Moammar Ghadafi.

Barack Obama won the Nobel Peace Prize in 2009, and less than two years later we woke up to the realization that he was assassinating people around the globe.

He's become the Assassin-in-Chief of the United States, violating U.S. and international law, killing whoever stands in the way of the global takeover by Obama and his colleagues. He does it through remote drone attacks, or by sending in Navy SEALs to do his dirty work, or through revolutionary surrogates.

When deposed Libyan leader Moammar Ghadafi was assassinated by Libyan rebels, Obama took credit for the killing. Ghadafi's convoy was attacked by NATO forces as it tried to escape from his hometown of Sirte. He was wounded in the NATO air strike but escaped briefly, hiding in a drainage culvert until rebel forces found him. NATO says, "We didn't know it was Ghadafi."

In his comments on Ghadafi's death, Obama never mentioned that the deposed Libyan dictator was captured alive or that he pleaded for his life to the bloodthirsty vermin who captured him. As he was pulled out of the storm drain where he had been hiding, so-called freedom fighters shouted one of the worst insults you can hurl at a Muslim: "You dog! You dog!" According to eyewitness accounts, Ghadafi tried to reason with his captors. He demanded his right to a trial as he asked them, "Do you know right from wrong?"

Their response was to beat Ghadafi to within an inch of his life before they shot him.[30]

The world cheers, Obama pats himself on the back, and not a single Republican condemns the brutality.

Ghadafi was a monster, but like every criminal under U.S. and international law, he deserved his day in court.

He was wanted by the International Criminal Court. Why wasn't he arrested, arraigned, and sent to The Hague for trial?

All of this leads me to believe that he was killed on purpose. Were there CIA operatives on the ground in Libya who were told to kill him if they got the chance?

The rebels beat the CIA to the punch.

Now we have to ask why they wanted Ghadafi dead.

Why did they not want a trial?

What did Ghadafi know about higher-ups in the U.S. government, the French government, the British government that they did not want to come out?

How complicit were officials of the U.S. and Western European governments in the evil deeds of Ghadafi, including the downing in 1988 of Pan Am flight 103?

Even more to the point: Where was the outcry in the United States against Ghadafi's murder?

Why didn't we hear the peaceniks and the anti-death-penalty people shouting that Ghadafi was put to death without due process?

I ask you this: Are we going to descend to the level of the junta in Argentina, where unmarked cars take Americans off the streets, throw black hoods over their heads, and "disappear" them?

This deplorable event—the assassination of Ghadafi—gets right to the point of the fears that I express in the title of this book: *Trickle Down Tyranny*. The murder of Ghadafi should send chills of fear down the spine of every American, regardless of political affiliation or orientation.

It forces us to ask, "Who's next?"

Yet few if any leftists said a single word about the execution-style killing of an unarmed Ghadafi while he was begging for his life. Certainly not the government mouthpieces on MSNBC—Mostly Snide Nonsense By Communists.

As I said on my show after Ghadafi was killed, "Here we have the U.S. media—who, I think, to the last empty skirt, opposed the death penalty—yet they are feasting on Ghadafi's corpse. . . . Every day they espouse sympathy for murderers and rapists on death row—no death penalty—and yet here they celebrate the execution of Ghadafi by the puppets of the U.S. government."

The assassination of Ghadafi without a single voice other than my own being raised in protest on that day had me so upset emotionally that I chose not to go on the air for my regular Friday broadcast the day after the event. I later found out that the UN actually raised some of the same questions, but I didn't hear a peep out of any U.S. politicians, out of our president—nobody spoke out.

Nobody but me was willing to stand up personally. Nobody took a moral stand on this.

Why are anti-death-penalty liberals, of all people, supporting the execution of a head of state?

What did our Secretary of State, Hillary Clinton, say?

She gloated over his murder.

Her words were, "We came, we saw, he died."

We now have Hillary Clinton acting out her violent, bloodthirsty inner self, modeled on the Charles Dickens character Madame Defarge from *A Tale of Two Cities*.

Clinton celebrated the execution of a world leader—someone who begged to be spared—without a trial.

To Clinton it's one big Fort Marcy Park.

Do you remember that one?

The death of Vince Foster?

Foster was a deputy White House counsel for Bill Clinton. He died under mysterious circumstances. His body was found in Fort Marcy Park, outside Washington, D.C., after he allegedly committed suicide.

Like Ghadafi, Foster knew too much. He couldn't be trusted to be left alive, so he had to be eliminated.

Ghadafi could not be trusted to face trial.

He knew too much.

So he was eliminated.

Did you hear a single American politician say, "He was unarmed. He was begging for his life?"

It was a mob execution. Chicago justice. Al Capone would have been proud. The Chicago thugs who played such an important part in getting Obama nominated and elected and in determining the path of his presidency are no doubt in their glory.

The $2 billion it cost us to eliminate Ghadafi must have been well worth it, but it brings up a larger, more important question: Why does Obama look the other way when despots like Ahmadinejad in Iran and Assad in Syria murder their own people, yet he brags and cheers when Ghadafi is taken out?

Why, all of a sudden, are liberals so full of themselves?

Why, when Obama invades a foreign country and assassinates leaders willy-nilly, is it OK all of a sudden? But when Bush tried to waterboard terrorists it was a violation of human rights?

In June 2011, the International Criminal Court issued arrest warrants for the Libyan dictator, his son Seif, and Ghadafi's security chief for their alleged crimes against humanity. That should have protected Ghadafi when he was captured. He should have been detained for trial.

Instead, he was a victim of the very war crime he himself was accused of.

Do you know what the Geneva Convention says about war crimes?

I'll tell you.

According to Marcel Ceccaldi, a French attorney who had worked for the Ghadafi regime, "The willful killing [of someone protected by the Geneva Convention] is defined as a war crime by Article 8 of the ICC's Rome Statute. Kadafi's homicide shows that the goal of [NATO] member states was not to protect civilians but to overthrow the regime." [31]

Ghadafi's family is suing NATO based on its violation of Ghadafi's rights.

Should they also be suing Barack Hussein Obama as an accessory to murder?

What is happening in the United States as atrocities are committed in the Middle East without our intervention is very disturbing to me.

It leads me to ask this question: Are some dictators more equal than others?

Why is Mahmoud Ahmadinejad—the Hitler of Iran—invited to dinner at Columbia University, but Ghadafi gets a bullet to the head?

I'll tell you why.

The bigger picture of Obama's campaign of terror in the Middle East is this: It is Barack Obama's lifetime anti-Zionism—his anti-Semitism— being played out on the world stage. He is imposing his personal hatred of Israel on the world order.

The George Soros-Jimmy Carter-Zbigniew Brzezinski-Barack Obama-Hillary Clinton complex is out to neutralize any Arab leader who is no threat to Israel and to support those who with no hesitation are in favor of the demise of the Zionist state.

Obama supported the arrest of Egypt's Hosni Mubarak—a friend of the United States, a friend of Israel for over 40 years. It was Mubarak who secured the border between Egypt and Israel during his time in office, who granted legitimacy to the state of Israel. Now Shariah law is being imposed in once-secular Egypt by the violent Muslim Brotherhood.

Obama has taken credit for killing Ghadafi, who, evil as he was, was not an enemy of the Jewish state as far as I know.

Notice who the U.S., under its anti-Zionist president Barack Obama, is *not* going after: Mahmoud Ahmadinejad, who threatens to wipe Israel off the map, and Bashar al-Assad of Syria, a puppet of Iran's Ahmadinejad, himself an anti-Israel despot who can be relied on to help carry out Ahmadinejad's crusade against Israel.

This is a massive anti-Zionist crusade. Obama is out to liberate Jerusalem for the Muslims.

His personality is merging with that of George Soros into a new being, a two-headed monster.

I call him *Sorobama*.

George Soros's fingerprints are all over Obama's foreign policy. This policy, as I explain in this chapter and the next, is in the process of completely realigning the power structure in the Middle East, making way for the emergence of a new caliphate, a coalition of Islamist governments determined to destroy Israel.

The *Sorobama* complex has targeted Israel, and the plan is being put in place stage by stage, killing or imprisoning leaders of countries that are neutral or friendly to Israel in the Middle East, and leaving Israel's archenemies alone.

Obama willfully delayed negotiations with Iraq on keeping our troops in that country and helping insure a necessary American presence in the Middle East. As a result of his intentional dithering, all American troops were to be withdrawn from Iraq at the end of 2011, ceding further power to Iran and other anti-American interests.

But as if Obama's withdrawal of U.S. troops from Iraq in order to reduce our influence in the Middle East and hasten the rise of an Iran-led coalition of Islamist governments in that region wasn't enough, he then bragged that the drawdown of troops in Iraq allowed the U.S. to kill Osama bin Laden.

While taking credit for the bin Laden killing, Obama neglected to mention that the killing of the al Qaeda leader was another of the growing number of illegal assassinations he has committed.

Like all leftist dictators, Obama is becoming a bloodthirsty monster.

Obama's greatest dream is to see Jerusalem divided. It would represent justification for his anti-Semitic Nobel Peace Prize, bestowed on him by the Jew-hating Norwegians.

Trickle down tyranny doesn't just mean guaranteeing loans to your political cronies at taxpayer expense.

It doesn't refer only to Obama's practice of supporting public employee unions in their quest to eliminate the private sector and make every American worker a government employee.

It doesn't just mean sending machine guns to Mexico without any arrests.

Do you realize that if a cop on the beat planted a weapon in a murder, that cop would go to jail as an accessory to murder? How are the president of the United States and his Attorney General any different from a cop who plants a weapon at a crime scene?

We're talking about the new world order of *Sorobama*, where George Soros is the puppetmaster and Obama is his puppet.

George Soros, the most power-mad man on the planet, a man who has more power than most sovereign nations.

Rothschild's quote applies here: "Give me control of a nation's money and I care not who makes its laws."

Soros has decimated the currencies of several nations—especially the British pound—and he's expanding his assault on freedom and capitalism through his control, not only of U.S. fiscal and financial policy, but of the American presidency.

The trickle down tyranny of the Obama administration is crippling our society and ushering in a new radical socialist era in which the rights of individuals disappear and the state assumes dominance over our lives.

It's the leftist dream becoming a reality.

It's America's worst nightmare.

Still have questions about Obama's position on the Middle East uprisings?

Let me make it clear: Obama paved the way for Islamist takeovers of the governments of former U.S. allies in the Middle East.

Those takeovers are continuing to happen as events move forward.

Bottom line: Ghadafi was a monster. But the Muslim Brotherhood is a more dangerous monster. Before they killed Ghadafi in cold blood, Libyan "revolutionaries" had repeatedly attacked Ghadafi's tribal homeland of Sirte. They were taking revenge on Ghadafi loyalists, terrorizing them and killing more than a thousand inhabitants of the region. One rebel fighter explained what was going on: "The Misurate brigades are taking their revenge for what soldiers originally from this village did to them. They are burning houses, stealing gold, and shooting animals."[32]

The United States had no vital national interest in Libya, yet we were willing to spend upwards of $100 million a day to keep Ghadafi's planes out of the air to help rebels who appear to be on the side of an Islamic government, many of whom are affiliated with Al Qaeda.

And what the world sees is the U.S. intervening in another oil-rich

Middle Eastern country. There was no chance that anything remotely re-sembling a democratic government would ever be formed in Libya. The in-nocents in the White House are projecting a cloistered, inexperienced view of the world onto a situation in Libya that promises only to move our coun-try further toward bankruptcy and the Middle East further toward Islamist domination.

There are no positives whatsoever in our intervention in Libya. We did the same thing in Egypt: We helped force out a U.S. ally and we made way for the Muslim Brotherhood to step in and take over the country.

Obama and the Destruction of the U.S. Military

Do you have any doubt that the Community Organizer-in-Chief is out to destroy the U.S. military?

The president is subverting the values and principles on which our coun-try was founded, on which we have risen to our position as the richest and most powerful nation in the world.

Nowhere is this clearer than when I examine what Obama is doing to undermine our military.

I've told you about how the administration conducted our involvement in the Libyan conflict and about how the Obama administration reacted to news of Ghadafi's death. Here are a few more things you need to know:

In August 2011, one of his staff sent more than 20 Navy SEALS to their deaths in Afghanistan on a slow-moving Chinook helicopter.

The president has not ordered an investigation of the incident.

As Commander-in-Chief, Barack Obama was responsible for the deaths of those Navy SEALS when they were sent to their death because they were put on a slow Chinook helicopter in order to transport them to a battle site! This was the single largest loss of SEALS in the history of this elite corps. Some in the military are whispering that this was an assassination of these fighting men.

To date, no investigation, no blame, no culpability!

Now, some would say, "What does this have to do with Obama?"

I wouldn't say that he had a direct hand in this, although some would say that as Commander-in-Chief he has direct responsibility.

However, let's ask this question: Why has there been no investigation? As a Commander-in-Chief who cared for his troops, he should be shouting every day for heads to roll in the military for the death of these SEALs, but we have not heard a peep out of the president about this so-called tragic accident.

Why has Obama said nothing about it?

Why has he buried this?

Even if he addressed this only for political reasons, he would at least be addressing it.

His silence on this matter shows his lack of integrity and morality. There should have been an immediate, full-scale, full-court press on the question of why these SEALs were put on that helicopter.

Why were they not sent in on the same advanced helicopters that our military used when they killed Osama bin Laden?

No one's asking that question.

They were sent in on a single transport helicopter that is known to be slow and vulnerable, a helicopter that was shot down by a single rocket-propelled grenade.

Who authorized the mission? Was it a general who issued the order? Worse yet, was it a girl from Harvard University who was put into the White House, like an agent of the KGB, who said, "Go ahead and use that helicopter?"

Or maybe it was just that the president couldn't keep his mouth shut. He had to take credit for the successful raid, and he identified to the world Navy SEAL Team Six so the Taliban knew who did it.

As I've explained to you, the president relies on three Ivy League-educated women to tell him how to run the military.

Maybe they decided that it should be done to save fuel and promote green energy.

We don't know why it was done.

The important thing is that Obama is Commander-in-Chief and he has not authorized an investigation, nor has he shown the correct moral outrage over the deaths of so many members of this elite fighting unit. This is one of the greatest travesties of his entire regime. That's why the buck stops in the Oval Office. It goes right to the president.

I believe that in the case of the deaths of the Navy SEALs Barack Obama may be guilty of undermining our military through his administration's direct intervention on the battlefield to aid the enemy in one of the most devastating blows our military has ever suffered. I am convinced that the crash in Afghanistan of one of our Chinook helicopters that resulted in the loss of more than 20 members of Navy SEAL Team Six in early August 2011,[33] was the work of an insider in the Obama administration who tipped off the Taliban on the path the flight would take.

The killing of the Navy SEALs demonstrates what happens when you put a subversive community organizer in charge of the most powerful military force in history.

The president's undermining of the military didn't stop there.

Are you aware of this incident? Forty-seven of our soldiers suffered head, spine, and pelvic injuries when more than 1,000 were injured, some seriously, parachuting in a mock battle scenario during training in Germany.[34]

Was Obama preparing us to fight World War II all over again?

The president has not ordered an investigation of the incident.

How about this one? The president managed to insure that nearly $600 billion will be automatically stripped from the U.S. defense budget when the "select committee" charged with finding $1.5 trillion in savings as part of the "debt ceiling" deal fails to do so.

It was set up from the start. All the Democrats have to do is refuse to go along with the legitimate cuts needed in the debt ceiling compromise bill, and the military budget gets cut so far that we won't be able to maintain our national defense.

Am I the only one who understands this?

Obama is using the debt ceiling "select committee"—which I call nothing more than a communist *junta*—to destroy the American military!

He made it clear two weeks before the November 23, 2011, deadline for the committee to come up with $1.5 trillion in cuts or else the military budget would all but disappear: He told the committee that he would not countenance their trying to stop the automatic budget cuts that he snuck past Republican legislators in the debt ceiling deal.[35]

In other words, our military is doomed.

How did Republicans let the president get away with this stealth undermining of our military, of our national security?

How is the most subversive president in American history still in office, despite the fact that he is clearly determined to reduce our military to second-class status?

Obama didn't stop there.

He deliberately tried to make sure one of the companies that his crony capitalist confederates invested in got federal funding, despite the fact that the technology they were developing interfered with the Global Positioning System satellites our troops use on the battlefield, compromising their ability to conduct operations.

I'm talking about the LightSquared scandal.

Are you aware of it?

In that one, Obama tried to convince four-star U.S. Air Force general William Shelton, who heads the Air Force Space Command, to change his testimony so that it would be more favorable to LightSquared, a company that was developing a broadband network service and trying to place it on frequencies right next to those of the GPS. The Pentagon had been worried since it heard about the project that it might interfere with military GPS.

When General Shelton gave a classified briefing to senators and congressional representatives and staff, he dropped a bombshell: *The White House had asked him to change his testimony so that it didn't jeopardize the Light-Squared project!*

The general was asked to falsely testify that the proposed LightSquared satellite broadband project would not interfere with U.S. military GPS technology. Never mind that the safety and security of our troops on the battlefield might be endangered—the important thing was that the LightSquared project be allowed to move forward.

Shelton refused to give in, instead testifying that the White House had tried to make him lie before a House Armed Services subcommittee hearing. In his testimony, Shelton said that to approve the project would be like letting LightSquared "put a rock band in the middle of [a] quiet neighborhood." [36]

More specifically, preliminary tests of the LightSquared product proposal indicated that even using a small portion of the band the company was granted a license to in 2004 would cause significant disruptions to GPS. CEO Sanjiv Ahuja complained that the company had been "forced" to spend upwards of $100 billion to figure out how to avoid the interference.

Let me fill you in on some of the details I've dug up.

The reason Shelton was told to change his testimony to favor Light-Squared? The company's majority owner is an investment fund run by Philip Falcone, and Falcone is a very big contributor to Democratic candidates. He and his wife gave more than $60,000 to the Democratic Senatorial Campaign Committee in 2009.

Falcone's story is important here, because it follows a common pattern of patronage and reward in the Democratic Party. Falcone grew up with eight other siblings in Chisholm, Minnesota. He was a good student and a hockey player, and he earned a scholarship to Harvard. After graduating with a degree in economics, he worked at the junk bond desk of Kidder, Peabody & Co. and spent time at Wachovia and Barclays as well. Like the other hedge fund managers who tanked the equities markets in 2008 and enabled Obama to be elected, Falcone bet against the markets and made a killing. He's worth about $2 billion and is number 540 on Forbes's list of billionaires.[37]

He's one of Obama's crony billionaires, and he expects to be treated like one.

It looks to me like Falcone and Ahuja felt they weren't getting their money's worth from the Obama administration. You see, Ahuja had committed a substantial amount of money to the Democratic Party in the past two years, too. LightSquared complained that the testing was taking too long, and in order to speed things up they tried to enlist Anthony Russo, director of the National Coordination Office for Space-Based Positioning, Navigation, and Timing, to insert the demand for a 90-day window for testing into his testimony before the House Science Committee. Russo refused, then compounded things by telling the committee what he'd been asked to do.[38]

Here's what he has said on the subject: "The GPS community is concerned because testing has shown that LightSquared's ground-based transmissions overpower the relatively weak GPS signal from space. Although LightSquared will operate in its own radio band, that band is so close to the GPS signals that most GPS devices pick up the stronger LightSquared signal and become overloaded or jammed."[39]

LightSquared had already more than doubled the size of its lobbying team by 2011, leading up to its confrontation with Congress and the Federal Communications Commission (FCC). It spent $830,000 in the first half of 2011 with the nine K Street lobbying firms it now employs. LightSquared executives have also been exchanging e-mails with White House aides in order to promote the company's interests.[40] The company's primary lobbying firm is owned by Norman Brownstein, one of Obama's biggest fund-raisers. In addition, on the advice of another of his big donors, George Haywood, Barack Obama at one time invested $90,000 in SkyTerra, the company that has since changed its name to LightSquared.[41]

Where there's the smell of a scandal, there's Obama's puppetmaster, George Soros. A group known as the Public Interest Spectrum Coalition filed a petition with the FCC in April 2010 in support of Harbinger Capital Partners in its attempts to enter into competition with AT&T and Verizon in the broadband network business. The Public Interest Spectrum Coalition is made up of four other groups—Free Press, Media Access Project, New America Foundation, and Public Knowledge. George Soros has contributed millions of dollars to these leftist activist groups: $1.4 million to Free Press, $1.7 million to Media Access Project, $3.8 million to New America Foundation, and $855,500 to Public Knowledge.[42]

Harbinger Capital Partners is Philip Falcone's company, the one that owns LightSquared. Soros Fund Management LLC is reported to have invested in Harbinger Capital Partners as well.[43]

Do you understand the extent of the corruption that this administration is involved in? Barack Obama was prepared to risk compromising a fundamental military system and the safety of American soldiers in order to protect his fund-raisers and promote their businesses.

U.S. soldiers' lives were put in jeopardy so Obama's cronies can continue to live the high life.

The U.S. military is being sacrificed on the altar of political donors.

I see Barack Obama's corruption and his support of the Islamist radicals at the heart of the so-called Arab Spring as amounting to war crimes. But they're not war crimes against our enemies. They're war crimes against the United States. They're war crimes against one of our most important allies, Israel.

What Obama and the left have done to undermine the U.S. military falls into the same category. Our armed forces have become a hotbed of political correctness toward gays, women, and Muslims that is rapidly destroying morale and fighting readiness.

Former U.S. Navy Secretary John Lehman, who was a carrier-based Navy pilot during his military career, didn't waste words in an article he wrote excoriating military officials for giving in to political correctness. In his words,

> The political correctness thought police, like Inspector Javert in "Les Miserables," are out to get [Navy pilots] and are relentless. . . . Those attributes of naval aviators—willingness to take intelligent calculated risk, self-confidence, even a certain swagger—that are invaluable in wartime are the very ones that make them particularly vulnerable in today's zero-tolerance Navy.[44]

Lehman explains that career military men now walk around in fear that they'll be reported for a homosexual slur or an "indecent" remark and be disciplined—or dismissed—for something unrelated to their military responsibilities, and the environment has become one in which the good of the country is subverted in the name of the good of minority groups.

The military carries its political correctness well beyond punishing what should be inoffensive behavior against social minorities, to failing to punish life-threatening behavior on the part of Muslim religious minorities. Even in the wake of Major Nidal Hasan's "alleged" murder of 13 military personnel in 2009 at Fort Hood, carried out in the name of Allah, the military remains afraid to stand up to this real existential threat. Another avowed Islamist militant, Private Naser Abdo, still faces trial for his attempts to purchase weapons that he intended to use in a Hasan-like attack. Prior to that, the army was weighing his request to be granted "conscientious objector" status for his Muslim religious beliefs, another way this administration grants validity to a religious belief whose core principle is the elimination of infidels.[45]

Since the ban on gays in the military has been lifted, the radical gay activist group OutServe is demanding that the armed forces target gays for recruitment in the military. For its part, the military has completed sexual "sensitivity training" that is the final step for the repeal of the gay ban.

What you don't know is that *the results of a study that indicated that the military was ready to let gays in were phony.*

The administration circulated the false information that 70 percent of military personnel surveyed felt there would be no problems connected with the repeal of the ban, in order to falsely convince senators and congressmen to agree to lifting the gay ban.

Elaine Donnelly, president of the Center for Military Readiness, found that exactly the opposite was true when the results were finally tabulated and published, long after the vote had been taken: "Nearly 60% of respondents in the Marine Corps and in Army combat arms said they believed there would be a negative impact on their unit's effectiveness in this context; among Marine combat arms the number was 67%."[46]

The U.S. Military Academy at West Point has been training and shaping the officer corps of the U.S. Army for more than 200 years. The stone and brick campus on the Hudson River does more than turn cadets into officers and produce men who lead other men into battle. West Point serves as a guardian of the army's most precious traditions and mentor to its highest-ranking generals until the day they retire. Throughout its long history, West Point has played a fundamental role in guiding the army as it evolved to meet new threats and adapted to new social realities. So positions on the U.S. Military Academy's Board of Visitors are highly prized and surprisingly powerful.

The Board of Visitors functions like a corporate board of directors that provides direction to the army's most sacred citadel. Under law, the U.S. Congress gets to appoint nine members of this powerful board and the president gets to appoint another six. So who did Obama appoint to this critical board at a time when the army is in combat in Afghanistan, Iraq, and the shores of Africa?

Obama appointed former Captain Brenda S. "Sue" Fulton, an openly gay former army officer. Fulton was not a believer in "don't ask, don't tell." She is a gay activist. She is a founding board member of OutServe, a group of active-duty military officers who are openly gay, and is a co-founder and executive director of Knights Out, an alumni group of gay West Point graduates.[47]

While she graduated from West Point in 1980 in the first class to include women, commanded a military intelligence unit, and was honorably discharged in 1985, Fulton has been a vigorous opponent of the U.S. Army's regulations and laws regarding homosexual conduct while in uniform. It is not her private life that concerns me here; she can do what she wants. It's that she went to war with army regulations year after year and is now in a position to implement a gay rights agenda in the military.

I don't care that she is gay, but I don't want her to turn our army into a social program at the same time it is trying to win wars on two continents.

Obama also appointed Patrick Murphy to West Point's Board of Visitors. Murphy, a former Democratic congressman and an Iraq War veteran, is a favorite guest of Rachel Maddow on her MSNBC show. What's his big issue? As a congressman from Pennsylvania, Murphy championed the House program for recruiting more gay soldiers into the U.S. Army.[48]

Whatever you think of the president's executive order and recent federal court cases requiring the military to open its doors to active gays and lesbians, designing, writing, and enforcing the rules of conduct for homosexual and heterosexual soldiers is going to be complicated and difficult. It is going to require balance and patience. Putting two gay activists into power means that the Board of Visitors will not command the trust of the troops or be able to debate the fine points of regulation with distance and sobriety. Instead, I expect Fulton and Murphy to be pounding the table for the most wild-eyed, politically correct rules possible. Soldiers in combat who use the wrong words might see themselves brought up on charges and their careers ruined. Subordinates who receive gay advances from their superiors will find a chain of command that looks at their plight with coldness and even hostility. Morale will suffer.

Obama just put the gay agenda on the table at West Point.[49]

Obama is also guilty of attempting to weaken our military by such actions as defunding the development of the F-35 fighter jet and shutting down our NASA space exploration program. As I've explained in chapter 4, he's banking on the debt ceiling junta not being able to find the necessary budget cuts so that the automatic cuts to the military of $600 billion kick in, guaranteeing that our military is further weakened.[50] He's bowed to Chinese pressure and agreed to cut weapons sales to our ally Taiwan at a loss of more than 16,000 jobs and nearly $800 million in revenues for the U.S.[51]

He's guilty of perpetuating the lie that there can actually be anything resembling "free and fair" elections in Middle Eastern countries when he knows, as the rest of us know, that a free election is an open invitation to the Muslim Brotherhood and other radical Islamist groups to intimidate the population and take over control of countries in turmoil. The invitation to hold "free elections" in Middle Eastern countries with no history of democracy and no democratic infrastructure or culture in place is nothing less than a naïve invitation to Islamist radicals to step in. The transition from a feudal Islamist theocracy to a modern democracy must be made very gradually, if it can be made at all. People with no history of establishing and maintaining democratic institutions must be led into their formation.

The Obama administration has no intention of leading that process. In fact, as I'll explain to you in the next chapter, Obama has reversed American foreign policy completely. Countries who have been our traditional allies we now treat as enemies, while those who have been our enemies are now our new friends.

CHAPTER 8

Tyranny of Treating
Our Friends Like Enemies and
Our Enemies Like Friends

In the 1950s, when I was growing up, they were called eggheads. Today they're called nerds. We're all familiar with them. They've got an excess of IQ and a shortage of testosterone. They spend most of their lives inside the academic incubator with no real-world experience. They've never met a payroll or responded to anything more urgent than a fire drill, and they've never dealt with anyone who challenged their leftist politics.

After they graduate, they shuffle from campus to Washington and back. They're our cabinet secretaries, ambassadors, and presidential advisors. They bring along a "globalist" perspective that is an absolute danger to the United States' position of economic and military supremacy in the world. Beatle John Lennon called them "college pudding." It's these people who advise Obama on how to conduct his foreign policy.

In the previous chapter I told you about how they decided which wars we would fight and which ones we would ignore based on the "responsibility to protect" doctrine. Now I'm going to tell you about how they decide which dictators to support and which to oust.

These Ivy Leaguers, especially President Obama, have trouble telling our friends from our enemies. They tend to get it all backwards—more *trickle down tyranny.* They ignore or attack our old friends and give aid and comfort to our former enemies and eventually hope to get them to switch places.

The "Soros Doctrine" in the Middle East

I can't count the number of times I've heard the phrase "international community" used by the Obama administration. The reason is that the term is code for "global government."

The intern*a-a-a*tional community. You have to make it sound like the bleating of a sheep when you say it, because that's how the people who are pushing for global government view everyone else on the planet. As sheep.

Global government is nothing more than warmed-over Marxist-Leninist communism on a planetary scale.

Did you think that what happened all over the Middle East during the "Arab Spring"—which has now become the "Arab Winter"—was somehow a spontaneous uprising of freedom-loving young people wearing blue jeans and carrying cell phones and determined to promote democratic government in their countries?

In fact, the whole situation has been carefully orchestrated from behind the scenes by George Soros, the same man who helped engineer Barack Obama's rise to power in the United States.

It starts with the fact that America is a rudderless, leaderless nation. Weak, inexperienced leaders like Barack Obama are easy prey for international power brokers like Soros. I explained in *Trickle Up Poverty* how Soros was instrumental in bringing about Obama's election by working to cause the financial meltdown that enabled Obama to win. Obama had been behind in all the polls up to the time of the September 2008, financial crash, but he never trailed after it.[1]

Soros's influence didn't stop with getting a malleable schoolboy elected president. It continued to significantly affect the president's foreign policy, from the fact that we now treat our friends like enemies and our enemies like friends, right down to how the administration conducted the war in Libya.

It means that everything is upside down and backwards in Obama's foreign policy, just like George Soros wants it to be.

I'll make it very simple for you. The Obama foreign policy can be summed up like this: Treat your sworn enemies like friends and your long-time friends like enemies. It's the Ivy League way.

When a genuine democratic uprising against an illegally elected Islamist government in Iran began in 2009, Obama stood mute as Iranians were murdered by Mahmoud Ahmadinejad's military in order to quell the uprising. Soviet dissident Natan Sharansky called Obama's inaction "maybe one of the biggest betrayals of people's freedom in modern history. . . . At the moment when millions were deciding whether to go to the barricades, the leader of the free world said 'For us, the most important thing is engagement with the regime, so we don't want a change of regime.' "[2]

After Syria's Iranian puppet dictator Bashar al-Assad's forces had attacked and killed dozens of unarmed citizens as they protested against his regime,[3] Hillary Clinton praised him as a "reformer."[4]

In other words, if you're an ally of Iran and committed to the destruction of Israel and to the establishment of a new caliphate in the Middle East headed by a nuclear Iran, you've got U.S. backing.

It didn't stop with Syria.

Tunisian dictator Zine El Abedine Ben Ali had throughout his 23-year dictatorial reign been an ally of the United States and an enemy of Islamic militant groups. Protesters forced him to step down in January 2011, despite the fact that he had said he would not run for another term in Tunisia's 2014 elections. Again, dozens of people were killed in clashes with police in Tunisia without our feeling the need to step in. As one journalist noted, "Not one word of condemnation, not one word of criticism, not one word urging restraint came from Barack Obama or Hilary Clinton as live ammunition was fired into crowds of unarmed men, women, and children."[5]

And not one word of support for Hosni Mubarak, a Middle Eastern ruler who had kept Islamist radicals under control, was ever uttered by anyone in the Obama administration.

From the shameful Middle Eastern apology tour early in his administration to his selective support of radical Islamist dictators, Obama has made it clear whose side he's on.

In the spring of 2011, I explained in an article I wrote and posted on my website exactly what would happen as a result of the uprisings across the Arab world: Radical Muslim governments, with the support of George Soros and other anti-Western ideologues, would emerge where leaders had been ousted.

As I predicted, that is happening.

The war in Libya and the U.S. decision to commit resources to the ouster of Libyan dictator Moammar Ghadafi came during a time when the Middle East was in chaos, with "Arab Spring" uprisings against dictatorial governments happening in nearly a dozen countries. Obama's response to the uprisings reflected a disturbing trend, a pattern that's becoming apparent in the selective way he goes after Middle Eastern tyrants.

Every single one of the dictators the Obama administration has actively targeted for overthrow has been a U.S. ally.

Every one of the dictators Obama has refused to confront is the head of an Islamist regime that is hostile to the United States.

Do you realize that Obama's sympathy with the Islamist cause is so strong and pervasive that Congress had to pass a law banning the administration from further dealings with groups known to have ties to Islamist terrorists? In the continuing resolution passed in November 2011 that extended the government's ability to keep spending, there was a paragraph that prohibited "any formal non-investigative cooperation with unindicted co-conspirators in terrorism cases."[6] It forbids Obama from any further association with Islamist radicals out to destroy America.

It's easy to understand why this was necessary. The Islamist sympathizer in the White House began to tip his hand as the chaos in the Middle East spread. Let's take a Savage tour through the world and meet our new friends and new enemies.

Egypt

Egyptian president Hosni Mubarak had long been a U.S. ally and had kept the radical Muslim Brotherhood out of power for more than 40 years prior to Obama's intervention. The president suddenly insisted at the beginning of February 2011 that a transition to *democratic* government in Egypt "must be meaningful, it must be peaceful, and it must begin now."[7]

With that absurd statement—the idea that transitions of power in the Middle East could be peaceful and could result in democratic governments—Obama revealed one of two things: Either he and his Ivory Tower advisors are hopelessly naïve regarding what's going on, or he is secretly on the side of Muslim radicals and believes that the overthrow of our allies will hasten their rise to power.

Less than two weeks after Obama made his mincing declaration that Mubarak must go, his Director of National Intelligence, James Clapper, claimed that the Muslim Brotherhood was "largely secular." Within hours, the administration refuted Clapper's absurd claim, reinforcing the fact that its foreign policy is chaotic at best, subversive at worst.[8]

In addition to revealing the administration's ineptitude, its positions with regard to Egypt helped reveal where Obama really stands: firmly against a strong ally of the United States and Israel. While the Mubarak regime was dictatorial, it did maintain order in the streets and suppress Islamist radicals. It also maintained peace with Israel.

Within a matter of weeks, Mubarak stepped down. As the left-wing press and the Obama administration were congratulating themselves on ousting Mubarak, and the Muslim Brotherhood was moving toward taking over Egypt,[9] the Egyptian military stepped in and took charge.

If you didn't see what was coming, let me explain it to you.

As I told you in the previous chapter, Obama is an avowed anti-Semite. He's pro-Muslim, anti-Judaism. He's also anti-Christian, and his ongoing attack on Christianity is not confined to the United States. His support of the ouster of Mubarak—under whom Coptic Christians were allowed to worship in peace—has led to the rise of violence toward Christians by Muslims.

In October 2011, 26 Coptic Christians were killed and hundreds more wounded in attacks by the Egyptian military. The attack occurred as Christian groups marched through Cairo in protest against the burnings of their churches. Egyptian Muslims pelted them with rocks as they moved along, and by the time they had reached their destination at a radio and TV broadcasting facility, the army started shooting into the crowd and trying to run over the protesters with their vehicles. Observers predicted that the event would cause a massive emigration of Christians from Egypt.[10]

Obama's response left me disgusted. In the wake of the murder of more than two dozen unarmed Christians by the Egyptian military, the president called on *Christians* to show restraint!

How were they supposed to do that? By allowing more of their brethren to be murdered by the military?

The president continued: "Now is the time for restraint on all sides so that Egyptians can move forward together to forge a strong and united Egypt." The loss of life was "tragic," but Christians need to put it behind them?[11]

No international sanction against the Egyptian military?

No condemnation of an obvious hate crime against Christians?

No withdrawing of U.S. foreign aid from Egypt?

Don't tell me Obama didn't know something like this would happen.

After Mubarak's ouster, the Egyptian military demonstrated that it was incapable of maintaining order. Reports began to emerge out of Egypt that indicated there were no police on the streets in Cairo and other cities. Coptic Christians, who make up about ten percent of Egypt's population, were clashing with Muslims, and the result was extensive casualties. Ambulances were nowhere to be seen, and the wounded were transported to medical facilities in garbage trucks. Roadblocks were frequently set up, not by the government, but by lawless thugs who stopped traffic and stole valuables from the occupants of the automobiles they detained. Without a functioning police force, vigilante groups sprang up, taking the law into their own hands.[12] Reports also surfaced that the Egyptian army was partnering with the Muslim Brotherhood to perform "virginity tests" on women who protested in Tahrir Square.[13]

So much for Obama's insistence that the transition "must be peaceful."

So much for a pro-Israel, anti–Muslim Brotherhood Egypt as an ally of the United States.

Egypt is under military rule with the ouster of Mubarak. Although there remain deep divisions between Islamists and those who favor a secular government, the overwhelming likelihood is that the Islamist Muslim Brotherhood will prevail. With Mubarak gone, the transition to either a military government or one founded on Islamic law is guaranteed.

Libya

When the popular revolt against Libyan dictator Moammar Ghadafi began, Obama didn't know what to do.

Many of strongest supporters—like Nation of Islam leader Louis Farrakhan and race hustler the Reverend Al Sharpton—were actually in favor of the dictator.

And Obama had made nice to the dictator from his first days in office.

It is time for a Savage history lesson, which includes the important points that the mainstream media leave out.

Abdel Basit al-Megrahi was a colonel in Libya's intelligence service when he approved the bombing of a packed passenger jet, Pan Am 103. It exploded over Lockerbie, Scotland, in 1988—killing all 270 people on board. Mothers, babies, tourists, students, and two retirees coming back from a trip that they had dreamed about for a lifetime.

Ultimately, Libya turned over the bomber. He was tried in a British court and jailed for life.

The military intervention in Libya had everything to do with Europe's need to keep Libyan oil flowing and nothing to do with defending Libyan citizens or promoting democracy.

Have you forgotten Britain's and Scotland's deal with Libya?

The one that led to Megrahi's release?

British intelligence had been in bed with the Libyan dictator throughout Prime Minister Tony Blair's tenure in office. The Blair government had offered British special forces to assist in training Ghadafi's Khamis Brigade, the brutal and vicious Libyan security force. They'd also disclosed to Ghadafi how one of their secret forces operated. When it came to Megrahi, though, Britain caved. Ghadafi threatened "dire consequences"—including harassing British nationals and canceling lucrative oil contracts with British companies BP and Shell, and suspending cooperation with British intelligence—if Megrahi was not freed and returned to his homeland.[14]

The Obama administration played a role in that, too.

It secretly approved the release of supposedly terminally ill Megrahi. They said Megrahi was dying and asked if he could spend his last few moments of life with his family. A Scottish doctor played along. Of course, the families of the Lockerbie victims would have liked a few more moments with their loved ones before they died in a plane crash at Megrahi's hands.

The real reason Megrahi was released was so that oil giant BP could negotiate a contract with Libya.

With Obama's approval secretly delivered through the U.S. ambassador in London, Megrahi was flown home.[15] Cheering crowds in Libya greeted his plane like he had just returned from landing on the moon or winning the World Series.

As for the terminally ill Megrahi, he was still alive almost two years later. The Libyans had tricked Obama. Maybe he wanted to be tricked.

Still, Obama treated Ghadafi as a friend.

In Libya, the Transnational National Council (TNC) has gained some control as the governing body of Libya and has been recognized by the U.S. and other European countries. The Muslim Brotherhood, banned by Ghadafi, would be eligible to participate in a government formed by the TNC. The U.S. went along, recognizing the "legitimacy" of the Brotherhood in Egypt in July 2011.[16]

Pakistan

Osama bin Laden was hiding out in a concrete castle some 800 yards from Pakistan's equivalent of West Point. The Pakistanis said they had no idea the arch-terrorist was there.

Barack Obama believed them. After all, the Pakistanis were our friends.

The question Obama failed to ask was this: Why wouldn't bin Laden be in Pakistan?

Nearly every senior al Qaeda leader that we've taken out has been captured or killed in that country. More than two-thirds of all al Qaeda "high-value targets" in the world have been killed or captured in Pakistan. That's more than were killed or captured in Iraq and Afghanistan combined.

Where was Ramzi Yousef, who bombed the World Trade Center in February 1993, captured? Pakistan.

Where was Khalid Sheikh Mohammed, the planner of the September 11 attacks, captured? Pakistan.

Where was Ramzi bin al-Shib, the so-called 20th hijacker, captured? Pakistan.

Where was Abu Zubaydah, a key al Qaeda supervisor linked to the 9/11 attacks, captured? Pakistan.

Where was Amar al-Baluchi, who carried money for the September 11 attacks, captured? Pakistan.

Where was Abu Faraj al-Libi, the head of al Qaeda's military wing, captured? Pakistan.

The last two heads of al Qaeda's military wing were also killed in Pakistan. The head of Al Qaeda's computer network was captured in Pakistan. So was bin Laden's doctor. So was the courier for Mullah Omar, the founder of the Taliban. So was Mir Aimal Kansi, who shot several CIA officers at the spy agency's front gate in 1993.

I could go on, but you get the point.

The Pakistanis are up to their eyeballs in senior al Qaeda figures—but they never seem to be able to spot them.

And virtually all of these terrorists were captured in Pakistan's major cities—not hiding in some distant mountain cave. Often they were found in Pakistan's capital city, Islamabad. So we're not talking about some terrorist who disappeared into the folds of the Hindu Kush Mountains. No. They were all found in large homes or pricey hotels in the wealthiest neighborhoods in Pakistan's largest cities. It would be like the Unabomber hiding out in Beverly Hills.

And it seems anyone in Pakistan who wanted to find bin Laden or a senior al Qaeda figure could do so. Hamid Mir, a reporter at *Dawn*, one of Pakistan's largest English-language newspapers, interviewed bin Laden more than any other journalist did. Each of those interviews was conducted in Pakistan. When Al Jazeera wanted to interview the mastermind of the September 11 attacks, Khalid Sheikh Mohammed, they found him in Pakistan. When *Newsweek* wanted to interview Taliban leaders who escaped from an Afghan prison, they did it in Pakistan.[17]

The only people who couldn't find bin Laden or his top henchmen were our allies, the Pakistanis.

When European intelligence services monitored phone calls from suspects in Europe in 2001, they found that more than 60 percent of all calls went to a single city in Pakistan, Karachi. The following year, more than half of all suspect phone calls went to another Pakistani city, near Pakistan's Jalozai refugee camp.

Meanwhile, President Obama asked Congress to send Pakistan another $3.4 billion in military and foreign aid in 2012. That was on top of the $4.46 billion we sent them in 2010. And the estimated $3 billion we sent them in 2009.[18]

That's the equivalent of sending everyone in Pakistan a check for almost $3,000 every year.

So after bin Laden was shot dead in his Pakistani hideout, what did our friends in Pakistan do? Did they apologize for missing the world's most wanted man in their midst? Did they promise to try harder?

What do you think?

Pakistan's President Asif Zardari took to the pages of the *Washington Post* to complain. How can the United States complain, he wondered, when Pakistan was doing the best it can?[19]

You can't say he doesn't have chutzpah.

Then our friends the Pakistanis warned us never to kill another terrorist in their country without telling them first.

Should we tell them?

Let's not forget what happened the last time we tipped off the Pakistanis that we were going to kill bin Laden. It was August 20, 1998. A few weeks earlier, bin Laden's men had driven truck bombs to the gates of two U.S. embassies in East Africa, killing 224 people (including 12 American diplomats). So, after many meetings with his cabinet and his national security team, President Clinton decided to launch dozens of Tomahawk missiles against bin Laden's mud-walled compound in Afghanistan. When the missiles were just about to cross over Pakistan's airspace to landlocked Afghanistan, a visiting U.S. general told the head of Pakistan's military that the missiles were flying through his country to hit bin Laden in Afghanistan. The Pakistan general asked if he could be excused for a minute to confer with his chain of command. Minutes later, U.S. satellites observed panic in bin Laden's compound and trucks full of people darting off in all directions. The missiles landed ten minutes later.[20]

Bin Laden got away clean. And soon he was planning the September 11, 2001, attacks.

Still, Obama talks about our good friends the Pakistanis.

And then Obama began acting like we were the problem.

Instead of calling on the Pakistanis to hand over al Qaeda operatives hiding in their country, Obama announced a multimillion-dollar "Muslim outreach" program. Our friends the Pakistanis just don't seem to understand us. Professor Obama is going to fix everything with a speech and a pile of money.

Our new exchange of friends for enemies doesn't stop at Middle Eastern countries. Caribbean and South American communists are also our new allies.

Cuba

Remember the Graham Greene novel *Our Man in Havana*? It's a funny story about a vacuum cleaner salesman who sells blueprints of his Hoover vacuums to British intelligence, who are convinced he has stumbled onto some secret weapons being developed by Cuban communist leaders. Pretty funny, in a dry British way.

Well, Americans now have a real man in Havana. His name is Alan Gross. Let me tell you his story.

Gross lived in suburban Maryland, near Washington, D.C. He devoted his life to doing humanitarian work around the world. He helped farmers in Azerbaijan and Bulgaria boost the yields of their crop fields—saving many from poverty and hunger. He worked in the poorest and most remote section of Pakistan to help the people there attract investors to a local mine. Now ore is coming out of the ground and producing jobs. For more than 25 years, he worked in Africa and the Caribbean, helping the poor. He had never been in trouble with the law.[21] When the U.S. Agency for International Development, or USAID, an organization that has provided economic and humanitarian aid to people around the world, approached Gross, he was open to their request to help them distribute computer equipment to the small and isolated Jewish community in Castro's Cuba.

Gross understood that many of the island's few remaining Jews were old, sick, and poor. Many of their relatives had fled to Florida. They were unable to visit their relatives—Castro doesn't let people leave his island prison— and their families were unable to visit them. Letters and phone calls were often intercepted. Many Cuban Jews had not seen their children in decades and had never seen their grandchildren.

Having computers and Internet access would enable them to e-mail their relatives, even see them on their computer screens. It would change their lives.

Gross couldn't refuse.

In the fall of 2009, Gross made several visits to Cuba on a tourist visa, bringing computer equipment with him and helping to set up an intranet so the Jews scattered across Cuba could talk to each other without the communist authorities listening.

In December, 2009, before he was able to make the next step—linking the Cuban Jews to the outside world—he was arrested. The Cuban authorities phoned his hotel room and asked him to come down to the lobby to pick up a message. Instead, the secret police picked him up.

For the next 14 months, he was held in a maximum-security prison. He kept asking what crime he had committed or what the charges were against him. There were none. They were simply holding him.

The little food he got in prison was tainted, and Gross became very ill. When Judy, his wife of 40 years, was finally allowed to see him, in July 2010, she was shocked. In seven months, he had lost 90 pounds. He wasn't heavy to begin with. Only when the Cuban authorities believed that he would die soon was he moved to a military prison hospital, where he received some help.

In February 2011, Gross's Cuban attorney came to visit him. He was surprised and grateful that the American was still alive. He said he had good news.

You will never guess what that "good news" was.

The Cuban communists had decided to put him on trial. He was charged with being a spy and committing "crimes against the state of Cuba."

Why was that good news?

There was a slim possibility that he might be found innocent and released. Could that happen? His Cuban attorney, perhaps wanting to be kind, said, "It could."

But it didn't. In a trial closed to the American press, the prosecutor said

Gross was a "mercenary" sent to Cuba to wage a "cyberwar" against the Cuban revolution. Ordinary Cubans are not allowed access to the Internet and giving it to them was a "counterrevolutionary" act.

The judge sentenced Gross to 15 years in prison. Gross is now 61 and will likely die before he is released. After his arrest, his 26-year-old daughter was diagnosed with cancer. He may never see her again.[22]

Where was Barack Obama when all this was going on?

Playing golf, going on vacations, and ignoring what was going on in Cuba.

Throughout Gross's long ordeal and travesty of justice, Obama didn't utter a single word in his support.

To do that would be violating two of Obama's strongest positions: his anti-Semitism and his support for communist dictators.

The only thing we heard from our government came from the press spokesman for the National Security Council, Tommy Vietor. Vietor complained mildly that the communist judge's ruling "adds another injustice to Alan Gross' ordeal."[23]

That's it? Where is the outrage against an American unjustly held?

Why does Obama and his team have so much compassion for the terrorists held in a U.S. Navy base on one side of Cuba and no compassion at all for an American held on the other side of the same island?

Let's cut to the chase: For Obama, Alan Gross is just a pawn in a bigger game. Obama doesn't want to help this innocent American, because he has another agenda. He wants to improve relations with Cuba and its cruel dictator brothers, Fidel and Raul Castro, and he wants to make sure he doesn't offend his Islamist "brothers" by supporting a Jew.

Obama has lifted restrictions on Americans' sending money to relatives in Cuba. He wants to eliminate travel restrictions to Cuba. And he eventually wants to lift the embargo on Cuba. He wants to the Cubans to be our new friends, even if Alan Gross is left to rot in its prisons and his American family is relegated to staring at an empty chair where Dad used to sit. Imagine the pain of that empty chair on Thanksgiving Day or his wedding anniversary night.

That's the Obama doctrine: Turn our old enemies into our new friends, no matter what the human cost.

Venezuela

The story goes like this: As the president was flying home from a trip to Spain in November 2010, he stopped by the cabin where members of the press are sequestered and made a joke. He explained that he was returning to Washington by way of Venezuela so he could pay a visit to his friend, Hugo Chavez. CBS reporter Mark Knoller tweeted this: "Obama joked he'd have AF-1 fly home via South America so he could see Hugo Chavez. Some joke." Ed Henry of CNN had this to say: "POTUS a little punchy—comes to press cabin on AF-1 and jokes he's stopping in South Am on way home from Europe to see Chavez. . . ."[24]

Chavez was fine with the president's impromptu plan. He responded that he would "embrace" the president and "eat socialist arepas" with him.[25]

Hundreds of people responded to news of the incident when it was posted on the website The Blaze. One commentator had this to say: "Mr Obama will try to take America down with him and turn us into a Soros communist puppet. I for one will not let her pass into the night without giving my all to keep her free." Another one wrote, "Who thinks he wasn't kidding? Sounds like he just wanted to so see his friend and dictator . . . for some pointers on fixing the next election (if he 'allows' it at all) or taking over another private enterprise. WAKE UP America!"[26]

Venezuelan president Hugo Chavez is a dictator in the mold of his mentor, Fidel Castro. He has seized and shut down virtually all of the independent newspapers[27] and his broadcasting Gestapo, Conatel, has closed more than 200 TV stations and 34 radio stations that opposed him.[28] When judges dare to defy him, he has them removed.[29] Chavez has also seized the assets of nearly two dozen American companies—including Cargill, Coca-Cola, Conoco Phillips, ExxonMobil, and McDonald's, among others.

Let me give you an example of how this works. On October 26, 2010, Chavez said in one of his speeches, "The expropriation is already ready of this glass company." Then he acted like he forgot the name of the American company he was about to seize. "What's it called? Owens-Illinois. Expropriate it."

Within hours, soldiers arrived at Owens-Illinois' Venezuelan plant. If

Owens-Illinois was part of one of your mutual funds, Chavez just reached into your retirement account and stole a piece of it. The company lost millions in market capitalization that day.[30]

Got any of those stocks in your retirement plan? Well, Chavez just stole from you. This isn't small change. The price tag for seizing land, farms, sugar mills, and industrial facilities totals at least $23.3 billion, according to Ecoanalitica, a Caracas-based economic research outfit. If one counts the value of deposits seized in banks, Ecoanalitica says the total loss is more than $27 billion.[31]

If all this sounds familiar, it's because Barack Obama is trying to do the same thing here in the U.S. In many cases, he's succeeded.

No wonder people didn't think the president was joking when he talked about detouring so he could visit his friend Hugo Chavez.

Let me explain it to you.

Obama has been trying to shut down Fox News and conservative radio talk shows since he took office, through his appointment of an FCC transition czar, Henry Rivera. Rivera is a radical leftist lawyer, a former FCC commissioner who's working to try to eliminate commercial talk radio.

What's Obama going to replace talk radio with? I'd say that his test of the nationwide Emergency Broadcast System on November 9, 2011, should give you a pretty good idea of Obama's idea of good radio programming.

Obama has already seized two private auto companies—General Motors and Chrysler—and turned those assets over to the labor unions. He's taken over the student loan business so he can exert near-total control over what students learn and how they pay for it. And he's taken over the health insurance industry.

Maybe Obama doesn't feel too outraged because he has acted just like Chavez. As one commentator put it, "Of course, the Obama administration has engaged in some nationalizations of its own, with little or no compensation paid. Just ask the Chrysler bondholders or the shareholders in what is now officially called 'old GM.' So perhaps the administration has little sympathy for the small investor."[32]

You don't think we're on our way to becoming Venezuela?

You're not worried that Hugo Chavez is one of Obama's "friends?"

So Chavez is making war on your retirement and what does Obama do?

Nothing. No speech, no trip to the United Nations, no warning that the U.S. will seize Venezuelan assets in America if he doesn't return our property in his country. He does absolutely nothing.

In fact, the Obama administration couldn't care less. When a reporter finally worked up the nerve to ask the then State Department spokesman, P. J. Crowley, about the theft of American property in Venezuela, here was Crowley's rambling response: "Well, you know—statements are one thing. We'll see what actual actions, you know, take place. But, you know, we would expect Venezuela to provide prompt, adequate, and effective compensation for any expropriation of the investments of Owens-Illinois in accordance with international law or any other, you know, private business doing—present in—in Venezuela."

The only other Obama administration official to even mention the pocketing of U.S. assets by foreign dictators was Assistant Secretary of Commerce Walter Bastian. Here's what he said: "We have no problem with expropriation [theft by a foreign government], as long as those affected are paid."[33]

No problem with theft? And Bastian certainly knows that Americans are not paid fair-market value for their holdings. They get whatever Chavez feels like paying that day. Most are not even paid at all.

Behind the scenes, Obama administration officials are telling executives not to worry. By being nice to Chavez and not protesting the thefts, they hope that Chavez will eventually see his way clear to pay the American investors something. These Obama wimps are like the kids on the playground who tell the victim not to report the bully to the teacher and not to fight back. Be nice, they coo, and he will be nice to you later. Well, it doesn't work in the schoolyard and it doesn't work in foreign affairs.

Obama's approach to Latin America is the opposite of Teddy Roosevelt's. Obama's motto: Speak softly and carry no stick.

Guess what? It hasn't worked.

ExxonMobil and ConocoPhillips, two big energy companies with millions of American investors, gave up on Obama. They have decided to go to the World Bank's Center for Settlements of Investment Disputes. They are seeking tens of billions of dollars in relief for their lost oil facilities in Venezuela. It will take years before they even get a hearing. And even if they win, Venezuela can always ignore the World Bank.[34]

When Chavez isn't stealing from Americans, he is impoverishing his own people—that's why they keep sneaking across our borders.

Everyone knew that Chavez's takeovers of private industry would make Venezuela poorer, hungrier, and less democratic. Today Venezuela has the highest inflation rate in South America. Once a food exporter, Venezuela now has to import more than 70 percent of its food and shortages are common. The little food available is at the end of the long line at a government-controlled store. Crooked government bureaucrats—Chavez has his capitalist cronies just like Obama does—make the shortages worse by selling food in the black market at huge profits for themselves.

What isn't lost to corruption is lost to government incompetence. Some 130,000 tons of food was left to rot on ships in Venezuela harbors because the bureaucrats could not arrange a profitable enough black-market deal for themselves.[35] When this was reported by one of the country's few remaining independent newspapers, Chavez was forced to arrest the head of his food police.

His replacement is just as corrupt, according to local papers.

Before Chavez, there was no black market in food. The stores were full of fresh produce and local meat. Now there are ration cards and sad walks home with no food to be found.

Why don't the people of Venezuela do something?

They're trying.

In the September 2010 parliamentary elections, voters overwhelmingly supported the opposition, which was called the Coalition for Democratic Unity. The coalition won half of the popular vote, and when they were sworn in on January 5, 2011, they controlled 40 percent of National Assembly seats, up from just four percent the year before. Other opposition parties grabbed another ten percent of the seats.

The opposition probably would have swept all electoral offices if it wasn't for Chavez's thugs—who might have taken lessons from the New Black Panther Party in Philadelphia during the 2008 elections here—openly intimidating voters and manufacturing fake ballots.

But still the people did the best they could and won working control of the national parliament.

What was Chavez's response? He stole the parliament's power.

Days before the opposition parties were set to be sworn in, Chavez asked the old parliament—where his United Socialist Party controlled 83 percent of the votes—to give him the dictatorial right to rule by decree until the 2012 presidential elections. They voted overwhelming in favor. Chavez only had to make announcements on television and they became law.

Forget elections, legislative debates. The maximum leader simply mumbles something on television and it is law.

And the parliament is powerless to stop him.

After he received this dictatorial power, he bragged to a state-run newspaper that the newly elected opposition was now powerless: "Let's see how they are going to make their laws now."[36]

With Chavez starving his country and mocking democracy, there can only be two results: more seizures of American property in Venezuela and more migrants slipping over our border.

Again Obama does nothing and says nothing.

State Department spokesman Crowley managed only a lukewarm response: "What he [Chavez] is doing here, we believe, is, you know, subverting the will of the Venezuela people."

And that's it. Except for that stray remark by Crowley, you will find no response by the Obama administration. Pore over the transcripts of the State Department briefings and read all of its reports. You will find nothing—not a single line—that is critical of Chavez's takeover of democracy in Venezuela itself.

Congresswoman Ileana Ros-Lehtinen, who is the chairwoman of the House Foreign Affairs Committee, sees right through Obama's game.

The president might act like he's not taking sides, but he is. And he is siding with a socialist dictator rather than Americans who are losing their retirement money and Venezuelans who are losing their freedoms. Ros-Lehtinen was dead on when she said, "Choosing to take no side in the battle between tyranny and democracy in Venezuela only helps the tyrannical side."[37]

The good news is that Chavez will almost certainly be dead before the next election. His cancer—the cancer he claims has been cured—is accelerating, and his medical team gives him no more than six months to live. Apparently the advanced treatment for the disease that he received in Cuba didn't do the trick.[38]

The only questions now are, "Will the Chavistas who support the dying dictator be able to hold on to the power and the wealth they have stolen from the Venezuelan people?" and "Will Barack Obama intervene on behalf of those people to insure that Venezuela becomes a free country again?"

My advice? Don't put any money on Obama coming through for the Venezuelan people.

Brazil

I mentioned the Obama trip to Brazil in the previous chapter. When President Obama got invited to a summit meeting in Brazil, he decided to take the whole family. A little vacation in the sunshine while icy rains fell on Washington in March 2011. Of course that little trip cost taxpayers tens of thousands of dollars.

Why not?

He felt completely at home with the Socialist Party leaders of that rainforest nation. They spend their days criticizing the United States in the halls of the United Nations and blaming America in the local newspapers for their problems.

Just like Obama.

Once there, Obama announced that U.S. taxpayers would loan the Brazilian oil giant Petrobras some $2 billion to drill for oil off their coast in the deep waters of the Atlantic. As I've told you, he backed that up with a promise that America would be Brazil's customer for the oil they produce.

Of course, the president has blocked American oil companies from drilling in a number of offshore areas and in many inland oil fields. Obama's drilling moratorium costs the American economy some 100,000 jobs and billions in revenue per year, according to the American Petroleum Institute. That might be the reason he promised that America would be Brazil's "best customer." When he got back to the states, Obama whined about America's dependence on "foreign oil imports"—the very ones he had just agreed to subsidize in Brazil with our tax dollars.

I have a lot more to say about Obama's energy edicts in chapter 9, "Tyranny of Green Energy."

Our New Enemies

Government brutality is a fact of life in the Middle East. It doesn't matter whether it's an Islamist dictator like Iran's Mahmoud Ahmadinejad or an old-fashioned third-world tyrant like Hosni Mubarak or Moammar Ghadafi. Middle Eastern countries have since World War II generally been held together by force. Now that the Obama administration has found a "principle"—"responsibility to protect"—to stand on, it has a rationale for extending its policy of reducing U.S. influence in the region by promoting the rise of Islamist governments "of the people."

It's the Soros Doctrine at work.

I'm the only one who saw what was really going on in the Middle East as Barack Obama turned a blind eye to the violence of dictators against their citizens and supported the overthrow of U.S. allies in order to pave the way for the takeover of their countries by Islamist radicals.

I'm the only one who understood that Obama was implementing his new foreign policy, the one that says, "Our former friends are now our enemies." It's the second part of Obama's new foreign policy doctrine.

Israel

I want you to take a look at what Obama's support for the Islamist radicals at the heart of the so-called Arab Spring has done to Israel.

In May 2011, the president delivered two speeches on Israel in which he called for Israel to restart negotiations with its Islamist enemies, starting with "borders . . . based on the 1967 lines with mutually agreed swaps."[39] In other words, Israel—which Obama continually refers to as an "occupying force"—would give up large amounts of its territory while accepting borders that would prevent it from defending itself against the attacks from its sworn enemies just across its borders.

Israeli Prime Minister Benjamin Netanyahu proceeded to scold our schoolboy president in a speech to a joint session of Congress, saying, "Jerusalem will never again be divided. Jerusalem must remain the united capital of Israel." He received 29 standing ovations during his congressional address.

One of the lines that received an ovation touched on something Obama seems unable to recognize: "In a region where women are stoned, gays are hanged, Christians are persecuted, Israel stands out. It is different."[40]

After that, Obama and Netanyahu met in the Oval Office, where the Israeli Prime Minister made Obama look like a foolish schoolboy. Netanyahu explained that peace on the terms Obama had laid out "will crash eventually on the rocks of Middle East reality."[41]

In the meantime, Israel's border with Egypt is no longer secure after Mubarak's ouster. In August 2011, a group of as many as 20 al Qaeda–affiliated terrorists crossed over the Egyptian border into Israel, firing mortars, RPGs, and handguns at a bus and automobiles inside Israel, killing seven Israelis.[42]

In the same month, demonstrators outside the Israeli Embassy in Cairo carried signs with swastikas on them, shouting, "The gas chambers are ready!" A month later, Egyptian protesters stormed the embassy, overwhelming security guards and forcing its evacuation.

Iranian defense minister Ahmad Vahidi spoke for Iran's growing number of allies in the new Middle East when he said that "given the takeover of the Zionist regime's espionage den and the escape of the ambassador of this usurper regime from Egypt, we can say that the second wave of Islamic awakening against the Zionist regime has started and we hope that his movement will continue with more power."[43]

At the same time, Palestinian leader Mahmoud Abbas petitioned the United Nations to recognize a Palestinian state. While the Obama administration sidestepped the issue of Palestinian statehood in its response, it again criticized Israel's construction of "settlements" in areas that the Palestinians claim for their sovereign state.[44] The Obama administration also failed to note that the Palestinian Authority chose the mother of four terrorist murderers, one of whom killed seven Israelis, to lead the procession to the UN offices in Ramallah. She would be the one to hand over the letter to UN Secretary General Ban Ki Moon.[45]

Iranian Revolutionary Guard official Masoud Jazayeri summed up the response to Israel's presence in this way: "The noose becomes tighter around the United States, the Zionist regime, and those who have infringed the rights of nations, and their allies will experience the same fate."[46]

Israel is the only democracy in the Middle East and our oldest, most dependable ally in the region. So, in Obama World, that makes it one of our enemies.

First the U.S. State Department actually listed Israel as a state sponsor of terrorism. They quickly backtracked, calling it a "mistake."

Right.

Of course, Obama has to be subtle in his opposition to Israel. An overwhelming number of Americans support the Jewish state.

In her position as UN Ambassador, Albright disciple Susan Rice finally got the chance to make it clear just how strong her and Obama's hatred of Israel is. After she was forced to veto a UN resolution declaring Jewish settlements in Judea and Samaria illegal, she went on to throw Israel under the bus, explaining that the administration continues "our opposition to continued settlements," that "we reject in the strongest terms the legitimacy of continued Israeli settlement activity."[47] Another Obama appointee, U.S. ambassador to Belguim Howard Gutman explained that we need to distinguish between legitimate anti-Semitism and hatred of Jews that is justified. In Gutman's warped view, traditional anti-Semitism on the part of Muslims is wrong and not to be tolerated. On the other hand, anti-Semitism resulting from Israel's treatment of Palestinians is just fine and should be encouraged.[48]

It didn't take long for the administration's anti-Semitic policies to surface right here in the U.S.

I explained the anti-Semitic bias of the Occupy Wall Street demonstrators in chapter 5. You don't think that Obama's support of the anti-Semites in the Occupy Wall Street demonstrations is an accident, do you?

Did you think he was simply unaware of the anti-Semitic bias of these punks?

What you're seeing emerge from the Occupy Wall Street idiots is Obama's new friends-to-enemies and enemies-to-friends policy. His treatment of Israel—shown in his refusal to denounce the anti-Semitism of the demonstrators—is just one more example of his insidious treatment of our former allies.

Great Britain

Barely anyone noticed that Obama was not invited to the royal wedding in the summer of 2011. But it was quite a snub.

American presidents are usually invited to royal weddings and coronations. President Reagan was invited to the wedding of Princess Diana and Prince Charles in 1982. President Eisenhower was invited to the coronation of Queen Elizabeth II in 1952. They might not attend, but Americans presidents are always invited.[49]

Not this time.

The president of Uganda got invited. So did the potentate of Bhutan. Even the head of St. Vincent, a tiny isle in the Caribbean, got the gold-leafed invitation.

But not the leader of the free world, the president of the only remaining superpower on planet Earth.

The "special relationship" with Great Britain, which was forged in the fires of World War II and has lasted through 11 presidents and more than 50 years of world crises, isn't so special anymore.

At least, it is not so special to President Obama.

For a long time, the British didn't get the joke. They still thought America cared about them. Let's not forget that the British have deployed soldiers, ships, and warplanes to Iraq and Afghanistan to fight alongside our forces—losing thousands of men in the bargain. Or that British intelligence supplies thousands of phone taps as well as human intelligence to stop al Qaeda terror attacks here in the United States. Or that the United Kingdom, the world's fifth-largest economy, is one of America's largest trading partners and a top-five source of foreign direct investment in America—building plants and creating jobs here in the land of ten percent unemployment.

What happened? From the start of his presidency, Obama signaled his anti-British stance. Remember all those remarks about the evils of British colonialism?

Remember when Obama ordered the return of the famous bust of Winston Churchill, Britain's heroic World War II leader. The lump of bronze was loaded with history and symbolism. Its return to Britain, without explanation, offended millions of Britons. The story and the snub played in the British press for weeks.

Next came Obama's odd, egotistical gifts to the Queen and the Prime Minister. Gifts are traditionally given during state visits. For the Queen, he gave an iPod loaded with Obama's speeches.

And to the Prime Minister, who had sent soldiers to fight alongside ours in Afghanistan and Iraq, he gave a collection of DVDs of American film classics like *Gone with the Wind.* But the DVDs don't work on British DVD players and, the prime minister carefully explained to British papers, he already had the good films and the bad ones he didn't want.[50]

Small wonder that when the Queen went to the Kentucky Derby last year, she didn't stop at the White House for a visit.

And, while the British are famous for their stiff upper lip, they decided not to invite the Obama clan to the royal wedding.

Honduras

Honduras has historically been one of America's most dependable allies in Central America. But when Honduran president Manuel Zelaya made a secret deal with Venezuelan dictator Hugo Chavez to overturn Honduras's democracy and make it a single-party communist state, Obama got worried. The problem was, he wasn't worried about seeing a democracy overthrown; he was worried about the rights of the communists trying to stage the coup.

As Zelaya's term was coming to an end in December 2009, he announced that he was going to stay in power. Emulating Chavez, he canceled elections. Then he shut down independent radio and television stations, threatened opposition parties, and brought in advisors recommended by Chavez and Fidel Castro.

Honduras is a country of some seven million people, and many of them realized what was going on. A new dictatorship was being born.

And they didn't want it. They had seen what Castro had done to Cuba and what Chavez had done to Venezuela. All through the spring of 2009, the tension built. Honduras realized that it was at a crossroads. The president of Honduras's senate called together the members of President Zelaya's own party, and they went to the Honduran Supreme Court and got an order to arrest Zelaya on June 28, 2009.

Zelaya was arrested and taken by helicopter to Costa Rica. Although Zelaya was dressed in a business suit when he was arrested, he changed into pajamas on the helicopter that took him into exile. After he landed, he gave a press conference in his pajamas, claiming armed men had awoken him in the presidential bed. From Costa Rica, Zelaya hired unemployed men in Tegucigalpa, Honduras's capital city, to stage demonstrations demanding his return to power.

Meanwhile, a transitional government reorganized the national senate and, with the approval of the supreme court, resumed governing the country.

As the Hondurans were fighting to save their democracy, what did Obama do?

First, he denied all members of the Honduran government visas to travel to the United States. The few Honduras government officials who were able to visit the United States and tell their side of the story had U.S. passports.

At the same time, the Obama administration demanded the return of ousted president Zelaya, the man who would be dictator.

Honduras's most powerful ally, the United States, had turned against it.

Obama cut off aid and loans to the impoverished country. Foreign investors were scared away. The White House called the protective and legal moves of the Honduras government a "coup" and encouraged the American media to criticize the country. Obama and the American media did everything they could to bring an acolyte of Chavez and Castro to power in Honduras.[51]

The Hondurans refused to budge. Legitimate elections were held on schedule and new leadership, under President Porfirio Lobo, was ushered in.

The communists—and Obama—lost. The greater loss, though, was of one of America's oldest friends in the region.

Even though the Obama administration did everything it could to make sure that a communist dictator would be able to remain in power, they were forced to retreat and approved the results of the election. Panama and Peru also recognized the legitimacy of Lobo's government. Likewise Canada and the European Union. The United States didn't reinstitute financial aid to the country, though.

The communist governments of Argentina, Brazil, Ecuador, and Venezuela did not recognize the election results, complaining that the elections were not "free and fair" like the ones Hugo Chavez holds.[52]

Friends vs. Enemies

Like the president himself, Obama's labor union allies are on the side of Islamic radicals in the Middle East and North Africa. In the midst of the transformation of the region into a permanent Arab Winter, Obama's supporters in the AFL-CIO are helping to fund what it calls the "popular mobilizations against corrupt, oppressive regimes in Tunisia, Egypt, Bahrain and throughout the Middle East and North Africa."[53]

But even worse than Obama's empty rhetoric are his actions in promoting the Muslim Brotherhood. He directed Frank Wisner, a former U.S. ambassador to Egypt, to meet secretly with Issam el-Erian, a senior leader of the Muslim Brotherhood, in order to discuss the fate of Egypt after Mubarak was deposed.[54] It is also reported that Obama himself met with members of the Muslim Brotherhood in 2009.[55]

In the face of Obama's dereliction of his duties as Commander-in-Chief of the U.S. military and his tacit approval of Islamist radicals in the Middle East, attacks on Israel from the Gaza Strip have increased dramatically.[56] Iran's Mahmoud Ahmadinejad, sensing U.S. sympathy for Islamist radicals and Obama's weakness, has declared that "[t]here is no room for [Israel] in the region."[57]

And since Obama abandoned Mubarak in Egypt, Saudi Arabia and the other members of the six-nation Gulf Cooperation Council have expressed their lack of trust in U.S. policy. They had encouraged Obama to support Mubarak, and in the wake of the president's defection, they're turning elsewhere for support against Islamists. It seems they don't trust an Obama foreign policy based on "universal values."[58]

With Mubarak gone, Iran now has another likely ally in Egypt. With Egypt no longer in its corner, Israel is now completely surrounded by radical Islamist regimes dedicated to its destruction.

The situation began to play out in September 2011, when protesters in Cairo went on a 13-hour rampage, storming the Israeli Embassy, destroying property in the building, and forcing the entire embassy staff to leave the country. Six Israeli guards were forced to take refuge inside a steel-doored safe during the attack.

It's something that would never have happened if Mubarak were still in power. Yet, in this case, security forces stood by, doing nothing as demonstrators tore down a concrete security wall around the compound. The ruling Egyptian military did not order police to stop the mob.[59]

Obama's anti-Mubarak position is already paying dividends for the Islamist Sympathizer–in-Chief as Israel's situation becomes more tenuous by the day.

The situation is ripe for Islamist terrorist organizations to move in and take further control of the Middle East.

Obama is their enabler.

Obama is the epitome of a weak liberal president. He mouths utopian leftist platitudes while the governments of our allies in the Middle East are challenged. Obama has actually said that the Muslim Brotherhood "must reject violence and recognize democratic goals."[60]

His apparent lack of understanding of the consequences of this type of rhetoric and of the positions he holds is unfathomable. It now seems more and more likely that at heart he is true to his Muslim upbringing and is bound to cede the power of the Judaeo-Christian West to Islamic tyrants.

He supports the demonstrators when there is a chance that his support will lead to the overthrow of our allies, but he keeps his mouth shut when Islamist regimes are threatened by popular uprising.

His weakness is having the effect of enabling a new caliphate to be formed in the Middle East. We're very likely witnessing the formation of a new Islamist alliance, led by Iran.

People who understand the true urgency of the situation have universally condemned Obama's positions. A piece titled "A Bullet in the Back from Uncle Sam" described "the politically correct diplomacy of American presidents throughout the generations" as "naïve" and quotes Israeli lawmaker Binyamin Ben-Eliezer as saying, "I don't think the Americans understand yet the disaster they have pushed the Middle East into."[61]

The coward in the White House waited until "the international community"—in the form of the UN Security Council—gave him cover by voting for sanctions against the Libyan dictator before he agreed that the United States would join in against Ghadafi.

Maybe you also noticed that China and Russia did not vote on that resolution. That's because China and Russia want nothing to do with global government. They understand the necessity of being strong nation-states.

They understand that if Barack Obama has his way, the United States will become just another mediocre nation on a planet full of mediocre nations, leaving the field open for them to step in and assume power.

They're watching, licking their chops, as he does just that.

As I've shown you, George Soros continues to exert a strong influence on the policies and pronouncements of Barack Obama and his administration. His influence includes promoting the Muslim Brotherhood to a position of power in Egypt.

The idea that somehow America is "responsible to protect" the rebels who ousted Ghadafi—although we ignored precisely the same situation in Iran in 2009 and in Syria as Assad murdered civilian protesters—seems to defy all logic, unless you look further to find the underlying doctrine for the administration's policies.

The decision to commit American military resources in Libya had nothing to do with America's vital interests. Defense Secretary Gates admitted as much. It had everything to do with the policies of George Soros and his hatred for Israel and for Western capitalist democracies.

Soros's purpose is to line his pockets and those of his cohorts through the formation of a global government, and the first step toward that is the elimination of dictatorial allies of the West and their replacement by radical Islamist governments.

As the Middle East is burning, the globalists are earning. George Soros and his cohorts use the upheaval in the Arab countries to line their pockets. Our president is an as yet unindicted co-conspirator.

CHAPTER 9

Tyranny of Green Energy

I am a staunch advocate for protecting our environment. I published my first book, *Plant a Tree: A working guide to regreening America*, in 1975, and in it I gave advice and know-how about the ways we can work to preserve and improve our planet. That advice is as valid and important today as it was when I wrote it. I continue to contribute to many organizations as committed to legitimate environmental causes as I am.

That's why I find this administration's corrupt, fascistic co-opting of the cause of environmentalism for the purpose of inserting the tentacles of government further and further into our lives among the most disturbing of all.

That's why I'm devoting an entire chapter of this book to showing you just how far Obama and his crooked cohorts have gone with their takeover of the cause of saving our environment.

Let me start with their attack on one of our most successful companies.

In 1986, the Gibson Guitar Corporation was struggling, about to go under. That was the year its current CEO, Henry Juszkiewicz, bought Gibson and through a brilliant marketing strategy turned it around, making it profitable within less than a year of taking ownership. Like me, Juszkiewicz is a committed environmentalist. He's a supporter of the Rainforest Alliance and the Environmental Defense Fund, among other environmental protection organizations.[1]

I'm still wondering what he did to deserve the treatment his company has received from the Obama administration.

On Wednesday, August 24, 2011, thugs from Eric Holder's Justice Department—armed to the teeth with automatic rifles as if they were shutting down a meth lab—raided the Gibson Guitar offices and manufacturing facilities in Memphis, Tennessee.[2]

They were there to seize wood the company used in manufacturing guitars, and company financial records, in order to determine if Gibson had been using wood that had been banned under a recently revised U.S. law.

The real purpose of the raid had nothing to do with rare wood: It was conducted to demonstrate the power of government regulators to impose their will on American businesses and, by extension, on the American people. They waved their guns in people's faces and threatened Gibson employees with jail terms before closing down the factory for the day and sending the workers home.

Juszkiewicz was clear about what he thought the Justice Department was doing: "Agents seized wood that was Forest Stewardship Council controlled. Gibson has a long history of supporting sustainable and responsible sources of wood and has worked diligently with entities such as the Rainforest Alliance and Greenpeace to secure FSC-certified supplies. The wood seized on August 24 satisfied FSC standards."[3]

I'm even more clear about what the Justice Department was doing: The raid was nothing more than the Justice Department attempting to intimidate Gibson without filing charges.

Gibson is one of the most successful musical instrument manufacturers in the United States. Its guitars have been used by virtually every jazz, blues, and rock musician from Charlie Christian to Albert King to Eric Clapton. Gibson created the first and most famous solid-body electric guitar, designed by and named for the legendary musician and innovator Les Paul.

They're also one of the most environmentally friendly companies I know of, committed to the conservation and wise use of the rare woods that go into some of their instruments.

As bad as the raid on the company was, the regulatory fascists aren't stopping at simply trying to shut down Gibson. They're after much more than that.

The musicians who own Gibson instruments that might be made from banned wood are subject to having them seized, even though most of the instruments in question were manufactured decades before the restrictions on the wood they contain were in place.

The raids were conducted under the Lacey Act of 1900, originally passed to regulate bird feathers used in ladies' hats. The law was expanded in 2008 to include wood imports.

The bans aren't just on the wood the bodies of the guitars are made of. They extend to fret boards, bridges, anything, no matter how small, that might possibly be made from a restricted wood at any time. Instruments that were made before the law was expanded in 2008 are apparently grandfathered in to the legislation, despite the fact that their owners and manufacturers could have had no idea they were playing or manufacturing anything illegal.[4]

It's more than simply protecting endangered species of plants and animals or preventing global warming, though. I've told you over and over that Obama's green energy initiatives are nothing more than dangerous power grabs designed to weaken the economy and give more control over our lives to the leftist elite.

Juszkiewicz makes it clear just how brutal the Holder/Obama storm troopers were and what he and his employees had to endure: "I was pretty upset. But you can only do so much when there's a gun in your face and it's the federal government."[5]

The federal government committed the equivalent of armed robbery from a company that is a symbol of American capitalist success.

From electric guitars to electric cars, the Obama administration is an administration of regulatory fascists determined to control or shut down everything from manufacturing to energy production.

The Environmental Nazi in the White House promotes these green energy initiatives, not because he's committed to saving the planet but because he's committed to taking it over.

The Politics of Green Energy

He's also committed to paying back his political supporters.

I've known for a long time that that's how politics works. When someone contributes money to your campaign, you find a way to repay them. There's no secret about that.

But when you combine political payback with a green energy initiative based on phony science and the investment of billions of taxpayer dollars in companies the government knew were about to fail, you get some of the biggest scandals in recent memory.

In 2009, the federal government guaranteed a loan of $535 million to a green energy company that had absolutely no prospects for success. Like much of the legislation this administration has produced, Obama rammed through the loan without the necessary due diligence and against the advice of the Office of Management and Budget.

The company was the poster child for the president's corrupt green energy initiative. Obama visited it in May 2010—well after his advisors were aware that the company was in trouble—promoting it as a company that would create green energy jobs. The White House went so far as to produce a video explaining how "companies like Solyndra are leading the way toward a brighter, more prosperous future."

There were some dissenters. In May 2010, one Obama fund-raiser wrote to Obama advisor Valerie Jarrett, "A number of us are concerned that the president is visiting Solyndra. Many of us believe the company's cost structure will make it difficult for them to survive long term. . . . I just want to help protect the president from anything that could result in negative or unfair press."[6]

By the fall of 2010, the administration knew that Solyndra was in trouble. Even though they continued to promote Solyndra as one of the success stories of the president's green energy strategy, the Department of Energy pushed hard for Solyndra to hold off announcing pending layoffs until November 3, 2010. E-mails reveal that one investor wrote this: "[T]hey did push very hard for us to hold our announcement of the consolidation to employees and vendors to Nov. 3rd—oddly they didn't give a reason for that date."[7]

The investor, Argonaut Private Equity, couldn't possibly have been that naïve, not with the midterm elections being held on November 2, the day before the date approved for a negative announcement.

Do they expect us to swallow the lie that delaying the Keystone XL pipeline until 2013 has nothing to do with the 2012 presidential elections, too?

This is the most corrupt, lying administration in the history of this country. Not a single decision is made without considering the Democratic far-left base or the fat cats and unions that hold this anti-American, anti-capitalist administration together.

The Panderer-in-Chief will stop at nothing in his assault on our freedom.

In promoting Solyndra, Obama was ignoring warning signs that had been surfacing for more than a year. A PricewaterhouseCoopers audit of the company two months earlier said that Solyndra "has suffered recurring losses from operations, negative cash flows since inception and has a net stockholders' deficit that, among other factors, raises substantial doubt about its ability to continue as a going concern."[8]

On top of that, the Energy Department's credit committee, which decides who gets federal energy money and who doesn't, had reached the conclusion that Solyndra's application for funds was "premature" in January 2009, just before Obama took office.

Inspector General Gregory Friedman, who was appointed to oversee the distribution of stimulus funds, said that the green energy projects being funded were so far divorced from the reality in the industry that it was like "attaching a lawn mower to a fire hydrant."[9]

After Obama was inaugurated, Solyndra funding was fast-tracked.

The reason?

His financial backers stood to lose big money.

Billionaire George Kaiser, a major Obama fund-raiser, was one of the early investors in Solyndra. With the company struggling, Kaiser and others ponied up an additional $75 million in financing early in 2011. In return, they received priority over other creditors—including the taxpayers guaranteeing the loan—to be repaid.

Another Obama fund-raiser, Steve Spinner, is a "top official" in the Energy Department charged with monitoring how guaranteed loans are made.

He ostensibly wasn't involved in the Solyndra decision, but not because there might have been any conflict of interest on *his* part: His wife's law firm had represented the company, so he stepped aside.

Do you have any doubt that the half billion in taxpayer dollars will be used primarily to make Obama cronies whole on this deal?

Or that the deal was rigged from the start?

Once it had secured the federally guaranteed loan, Solyndra went on a lavish spending spree. The company doled out $344 million on a new corporate headquarters building with a state-of-the-art conference room whose glass windows smoked over at the touch of a switch in order to conceal who was in the room. Just down the road, at the company's manufacturing facility, unsold solar panels piled up.[10]

Solyndra executives lied to the White House in May 2011, saying "we have good market momentum, the factory is ramping and our plan puts [us] at cash positive later this year," and two months later followed up with "We can assure you we have a path to profitability."[11] In a move that was a sign they knew Solyndra was in trouble, the Energy Department hired the investment bank Lazard Ltd. in mid-August for $1.1 million to give advice on how the company could restructure its finances "both in and out of bankruptcy," according to records.[12]

On April 15, 2011, and again on July 8, more than a dozen Solyndra senior executives gave themselves big cash bonuses that ranged from $37,000 to $60,000 apiece.[13]

One of the big shots at the foundering company, Karen Alter, Solyndra's senior marketing vice president, took home two $55,000 bonus checks. Alter had contributed $6,300 to Barack Obama and two other Democratic candidates. Ben Bierman, executive vice president of operations, was another big Democratic contributor. He gave Obama $2,300 and several other Democrats a total of $7,400. Both of these scammers received salaries of at least $250,000 in addition to their bonuses.

But those weren't the worst offenses. On July 1, 2011, Chris Gronet, one of the Solyndra founders, was "transitioned to the role of adviser and consultant" from his job as CEO, shortly before the company declared bankruptcy. His reward for running Solyndra into the ground? *A severance package worth nearly half a billion dollars!*[14]

In late August, less than a month after the CEO jumped ship, Solyndra laid off 90 percent of its workforce.

Less than two weeks after that, the company declared Chapter 11 bankruptcy.

Within days, Eric Holder's FBI agents raided Solyndra headquarters and seized company documents and computers.

A criminal investigation into Solyndra is under way and company executives might face criminal charges.

The company's AWOL CEO and its CFO were called before the House Energy and Commerce Committee to testify on September 23, 2011. Their response was to take the Fifth.

Because they're the targets of a criminal investigation, they can invoke their Fifth Amendment constitutional right to avoid self-incrimination. The fact that Holder had initiated a criminal investigation gave them the cover they needed to clam up.

I'm the only one telling you that *if Holder hadn't acted as he did, Solyndra execs couldn't have taken the Fifth and would have had to reveal what they knew about the scam or risk being in contempt of Congress.*

Holder's actions may well have been taken in order to prevent them from having to testify before Congress.

They might keep the White House from having to testify as well. E-mails indicate that the West Wing was monitoring the situation from the start and knew the risks it faced but went ahead anyway in order to protect Kaiser and other politically connected investors.[15] That's illegal, and should result in criminal indictments against anyone from the White House who was involved, including the president.

The scandal is compounded by the fact that the Obama administration rigged the process so that investors in Solyndra—who were also big Obama donors—would get paid back before the taxpayers who guaranteed the loan.

According to the Energy Policy Act of 2005, a 551-page legislative behemoth, federally guaranteed loans "have an assured revenue stream that covers project capital and operating costs (including servicing all debt obligations covered by the guarantee) that is approved by the Secretary and the relevant State public utility commission."[16] In addition, guaranteed loans "shall be subject to the condition that the obligation is not subordinate to other financing." [17] In other words, taxpayers are the first to get paid back in the event the loan goes bad.

There's no question that the Department of Energy broke the law when it restructured the loan agreement in order to give private investors precedence over taxpayers in the case of the company's collapse—which collapse the DOE knew was a certainty at the time the loan was approved. That didn't stop the government from doing what it did to taxpayers in the GM/Chrysler bailouts. Susan Richardson, chief counsel of the DOE loans program, decided differently. She explained that paying back taxpayers first was not "a continuing obligation or restriction" but applied only to the initial issuance of the loan.

She's saying, "The hell with the taxpayers the minute after the loan has been granted." [18]

Obama justifies throwing money away on high-risk, guaranteed-to-fail green energy companies on the grounds that China is doing it. That's his rationale for giving a company with revenues of less than $100 million annually more that half a billion taxpayer dollars.[19]

In fact, of the $20 billion in stimulus funds available for the DOE to use in the guaranteed loan program, $16 billion of it went to Obama contributors.[20]

Despite the fact that some e-mails related to the Solyndra scandal have been released, the White House is slow-walking its compliance with a subpoena issued by the House Energy and Commerce Committee in early November 2011.[21]

It's another example of the stonewalling that has been the hallmark of the leftist vermin in this rogue administration every time they're confronted with the reality of their corruption.

The Democratic donors who invested in Solyndra aren't the only ones making a killing. Lawyers connected with the Obama administration are going to feast on Solyndra fees, too. Former Massachusetts Governor William Weld, a Republican who supported Obama, is slated to make $825 an hour to consult with Solyndra during bankruptcy proceedings. Three others from the McDermott Will & Emery law firm will bill the government between $425 and $775 an hour to represent Solyndra.[22]

I want you to get an idea of how widespread this scandal was, and how deeply it went into the ranks of Obama's political and financial cronies.

Did you know that Nancy Pelosi was directly involved in this scandal, too?

Here are the details: Pelosi's brother-in-law received most of the last-minute $737 million guaranteed loan to Tonopah Solar Energy. That company is a subsidiary of another company, SolarReserve, which is building a large solar-thermal plant in the Nevada desert. Ronald Pelosi, Nancy Pelosi's husband's brother, is a vice president of the private equity company that is the primary funder of this project and which had a hand in Solyndra as well.[23]

Another famous political name cropped up in the list of cronies Obama gave money to for non-viable green energy projects: Robert F. Kennedy Jr. BrightSource Energy listed as its principal private investor a company called VantagePoint. RFK Jr. is one of the "venture partners" of that company. BrightSource listed more than $200 million in losses in its prospectus, and admitted that its "proprietary technology has a limited history and may perform below expectations when implemented."[24]

That was all the Obama administration needed to hear. It greenlighted the foundering company for $1.4 billion in taxpayer-funded bailout money.

If you think the FBI was interested in gathering evidence to prosecute Solyndra—or any other of the phony companies that have received tens of billions of our money—let me set you straight: If so much as a single Solyndra executive or any government official involved in the Solyndra deal is brought up on criminal charges, it will be because Eric Holder has been replaced as attorney general.

I'm telling you that there's no chance Holder's FBI seized evidence in order to make it public. I believe that Holder's ultimate purpose was to make sure the evidence in the case never sees the light of day. I'm convinced that the crooked Attorney General will do everything in his power to protect Obama's interests and those of his financial backers.

As I see it, Obama—because he's unable to understand anything but the socialist economic theory that was drummed into his head during his school years—has once again ignored the markets in order to impose his political agenda on American business and to protect his backers. Lifetime federal government employees have no place being venture socialists with other people's money. If they were in the private sector, they would have been out of business long ago, but because they're protected by taxpayer money, when they lose they simply move on to other projects and leave it to the lawyers to clean up. With our money.

This administration has also thrown away our money in other federally funded green energy projects. Beacon Power Corporation, which had received $43 million in loan guarantees from the same source that backed Solyndra, declared Chapter 11 bankruptcy in late October 2011. Beacon develops energy storage systems based on "flywheel" technology. The money was supposed to go toward building an energy storage plant in Stephentown, New York.[25] Flywheel energy storage technology is supposedly an alternative to chemical batteries for storing electrical energy and making it available in times of energy shortages. In the case of Beacon, two of their flywheel energy storage systems failed, one resulting in a "catastrophic" steam explosion at the company's Stephentown facility.[26]

Federally funded energy storage projects took another hit when a company called Ener1 was delisted from by NASDAQ after its stock price underwent a calamitous plunge because the company released false financial information. A class-action lawsuit against the company filed on behalf of shareholders claimed that Ener1 made "materially false and misleading statements concerning Ener1's financial condition and prospects." The company received $118.5 million in DOE grant money.[27]

There's more. The government also lost millions in funding these projects: $5.3 million to Evergreen Solar Inc., $500,000 to Spectrawatt (which, as it reorganizes under Chapter 11 bankruptcy, plans to move its manufacturing plant to China), $424,000 to Mountain Plaza Inc.—*after the company had declared bankruptcy*—for "truck stop electrification" so long-haul drivers could turn their rigs' diesel engines off during extended rest periods, and $10 million to Olsen's Crop Services and Olsen's Mills Acquisition Company, again after they had gone into Chapter 11.[28]

Another taxpayer-funded project, this one in Vice President Joe Biden's home state of Delaware, threatens to drive energy prices for that state's residents through the roof. Delaware legislators have put their constituents on the hook for underwriting the bad business practices of another energy company, Nichols & Driessen.

The legislators have committed state taxpayers to provide financial guarantees to a California company, Bloom Energy, that manufactures fuel cells that are supposed to replace natural gas and coal to power electric generating plants. No private investors could be found to back this loser of a green energy company, so the pols stepped in, guaranteeing the company's finances and huge increases—$600 million over the next 20 years—in Delaware residents' energy costs.[29]

It's not just the guarantee that I'm concerned with. The way the legislation is written, if the Delaware legislature in any way changes the law that requires Delaware residents to pay the energy tariff to underwrite Bloom, the entire $600 million sum will be due and payable on demand.

Bloom's fuel cells don't even represent clean energy. They use large amounts of two rare earth elements that are mined only in China, and China is controlling their distribution and cost. The United States imports 100 percent of these elements. On top of that, the company is privately held and is not required to share its financial data with the public.

Even ten years ago I wouldn't have believed that things could come to this, where a bunch of communist punks in our own federal government has become the controller of capital markets and is making lousy bets on companies that have no chance of success—and using our money to do it.

Did you know that the green energy funding scandals I've told you about are only the tip of the iceberg?

> The Inspector General appointed to look into the green energy funding scandals has opened up more than 100 criminal investigations into the Obama administration's illegal use of taxpayer money to fund projects for his cronies![30]

The economic illiterates in the Obama administration are trying to pick winners based on political favoritism. They're putting taxpayer money at risk, wasting money on paybacks to their fund-raisers that could be used in the market as legitimate venture capital for companies that might actually have a chance of survival and profitability. They're taking private capital out of circulation and using it to back guarantees to their cronies.

The Solyndra failure is important because it helps reveal the extent of the corruption that is rife in the Obama administration, but it's just the beginning of the president's plan to gain control over energy production and use that control to paralyze our economy.

I'm the only one who will tell you what's really happening in this area.

Obama's Energy Strategy: Beg, Tax, and Persecute

Obama's Secretary of Energy, Steven Chu, has made the president's policy on oil production very clear: "Somehow we have to figure out how to boost the price of gasoline to the levels in Europe." [31]

Chu's reason for this? The higher the price of gasoline, the more likely "green" energy projects like Solyndra—of which Chu was an outspoken supporter—are to be competitive in the marketplace. Without astronomical gasoline prices, green energy is dead.

Average Americans are the target of this strategy: If they're hurt financially by rising prices at the pump, they'll see the wisdom of going green.

Obama appointed Chu because he's a global warming propagandist of the first order. It's hard to believe someone with Chu's credentials—Nobel Prize-winning atomic physicist, head of the Lawrence Berkeley National Laboratory at the University of California, Berkeley—would stoop to using phony evidence and bad science to support his position, but like everyone in the Obama administration, Chu is shameless in lying to promote global warming: "Stronger storms, shrinking glaciers and winter snowpack, prolonged droughts and rising sea levels are raising the specter of global food and water shortages. The ominous signs of climate change we see today are a warning of dire economic and social consequences for us all, but especially for the poor of the world." [32]

The administration's strategy for raising prices is working.

At the end of 2008, gas prices in the U.S. averaged $1.61 a gallon. By late spring 2011, the price of a gallon of regular had reached $3.89. [33]

The public's dissatisfaction with high prices at the pump caused Obama to implement a three-step strategy.

He said it was to reduce the price of oil.

I say it was just the opposite: He needed to divert attention from everything he's doing to keep oil prices high.

First, he implored oil-producing countries like Saudi Arabia, telling them that they "need to increase supplies." [34]

His reasoning?

Out-of-control oil prices would be detrimental to the *global* economy.

In his own words, "We are in a lot of conversations with the major oil producers like Saudi Arabia to let them know that it's not going to be good for them if our economy is hobbled because of high oil prices." [35]

In other words, it was the fault of the Arab nations, and they were "hobbling" our economy by not producing more oil themselves.

After that, he threatened the oil companies because they make too much money, telling us that the tax credits given to oil companies for a variety of purposes, including to help them defray the expense of exploration, should be taken away.

The total revenue that would generate? Four billion dollars.

Obama explained that the United States could put the $4 billion we give to oil companies to better use. He must have been thinking that we could use the $4 billion to pay one day's interest on the $14 trillion national debt. He's rung up more than 20 percent of that debt in his first two years in office. [36]

Finally, at about the same time, in a town hall meeting in Reno, Nevada, the president explained that he had directed Attorney General Eric Holder to assemble a blue-ribbon panel to "stop oil marketing fraud." Citing speculators who buy and sell oil futures as another key reason oil and gas prices continue to rise, Obama said he wanted to make sure that "no one is taking advantage of the American people for their own short-term gain." [37]

That's it.

That's all the rookie in the White House was able to come up with.

Beg the Arab countries to ramp up production.

Further hamper American oil companies' ability to produce oil by removing exploration tax credits.

Investigate commodities futures traders.

I'll tell you what Obama didn't tell you about the oil industry and his own strategy.

The oil industry isn't the guilty party.

When you look more closely at a typical large oil company, you discover it's not their fault.

ExxonMobil is a perfect example of the fact that it's Obama himself who's committing the fraud. And he's doing it at the expense of the U.S. oil and gas industry and the American people.

We know the president is beholden to radical environmentalists. They're an important part of his base. I'll explain later in this chapter that he's appointed one of them to head the EPA, and I'll lay out the damage that agency is doing to our economy and to the U.S. energy industry.

It's because of his commitment to a radical agenda that Obama must make it appear that the production and use of oil and natural gas are so harmful to the environment that they need to be curtailed, even shut down.

To do that, he tries to convince Americans that the oil companies are making outrageous profits.

When Obama cites numbers like Exxon's $10.8 billion in profits for the first quarter of 2011, he means to shock you into thinking that the oil company is ripping you off.

Here's the truth: While $10.8 billion is a fairly large number, it pales in comparison with the president's own multi-trillion-dollar unpaid-for budgets. And it isn't the number you should be looking at in the first place.

ExxonMobil's gross revenues exceeded $114 billion in the first quarter of 2011. The company's profits for that period were $10.8 billion. That's a profit margin of a little over 9 percent, respectable but nowhere near excessive.

What Obama fails to note is that Exxon's pretax earnings were $18.9 billion, and that the company paid $8 billion of that amount in corporate income taxes for the first three months of 2011. That's a tax-on-income rate of 42.3 percent. High corporate tax rates such as this are one of the reasons many multinational corporations are moving their headquarters out of the United States to other countries, where corporate tax rates are more reasonable. The company also paid $10.3 billion in sales taxes and another $10.3 billion in other taxes, mostly property taxes for the land it owns in the U.S.[38]

Exxon sells only about three percent of the total gasoline and diesel fuel it produces in the United States. The rest is sold internationally.[39]

For every gallon of gas Exxon sells in our country, the company makes a profit of about two cents.

On that same gallon of gas, the U.S. and state governments earn between 40 and 60 cents through taxes levied on gasoline and diesel fuel.

Exxon's profit margin, again, is about nine percent. The government's profit margin is 100 percent.[40]

Exxon is a very large corporation. At the end of 2009, the company operated nearly 14,000 oil and gas wells in dozens of countries on five continents.[41] It's part of a very large industry, arguably the most important in the world.

And while large oil companies are very successful, making billions of dollars in annual profits, their profit margins are not large. In fact, the energy industry as a whole is not even in the top 100 for all industries where profitability is concerned. The 6.1 percent *average* profit margin for all the energy companies in the "Major Integrated Oil and Gas" category ranked 112th among all industries for the first quarter of 2011. Oil industry profits were far behind such other industries as Internet providers (23 percent), cigarette companies (21 percent), and magazine and periodical publishers (51 percent).[42]

Why isn't Obama going after these industries for excessive profits?

I'll tell you why: Because they don't get in the way of his design to nationalize the oil industry in the same way he's nationalizing the auto and health-care industries.

Don't believe me?

Are you aware of the president's latest attempt to cripple the oil industry and keep us at the mercy of the Islamist Middle Eastern dictatorships that now control our oil supply?

Have you heard about the Keystone XL pipeline?

It's a $7 billion project developed by a company called TransCanada to build a 1,700-mile pipeline to carry crude oil extracted from Canadian oil sands fields down to U.S. refineries on the Gulf coast. The company has already spent more than $2 billion on the steel needed to build the pipeline, and it expected approval after an extensive review process that had satisfied 57 specific environmental and other requirements and that exceeded all safety standards.[43] The U.S. Chamber of Commerce puts the number of American jobs that would be created by the project at "more than 20,000." That's not counting the thousands of other jobs and businesses—from restaurants to manufacturers—that would benefit from the project.[44]

Not a single cross-border pipeline request has ever been denied.[45]

The decision to allow the pipeline to go forward should have been automatic. Despite Obama promising to be Brazil's "best customer" for oil, that country's ability to produce it is declining. Right now, we're Hugo Chavez's "best customer." We buy 900,000 barrels of oil every year from the Venezuelan dictator;[46] the Keystone XL pipeline would pump that amount of oil to Gulf coast refineries. The Canadian pipeline would not only increase America's oil supply from a reliable source rather than from a South American tyrant bent on our destruction. it would mean that U.S. refineries would have additional work and that we'd be able to ship some of the refined oil back to Canada.

On Veterans' Day, 2011, Obama vetoed the project by delaying it until after the 2012 elections. He ostensibly bowed to the radical environmentalists who have impeded energy development in the U.S. for decades, saying that more study needed to be done to determine the impact on the environment and the climate. In the process, he went against the construction workers' unions, who overwhelmingly favor going ahead with the project.

In fact, he's just making sure we don't produce enough energy to maintain our status as the world's most powerful nation.

It's unbelievable to me that the American people aren't demonstrating in front of the White House demanding that Obama resign. This action alone is enough to demonstrate that he is not intelligent enough or informed enough to be our president, and that he has everything but the best interests of the United States in mind when he makes decisions like this.

It may well be that Obama just killed the Keystone XL pipeline project permanently by this delay. If Canada can't wait another year and a half to see if the pipeline will be built, there's a waiting buyer for their oil. All they have to do is transport it to British Columbia and load it on oil tankers for shipment to China.

That may be what the Traitor-in-Chief had in mind all along.

It's not just a sop to environmentalists. It's another direct attack on America—like the others I've been cataloging for you in this book—that Barack Obama has been waging since he came to power.

What no one seems to be paying attention to is the fact that Mexico is drilling a deepwater oil well in the Gulf of Mexico, only 22 miles from U.S. territorial waters. Cuba is drilling a deepwater well 60 miles from Key West, Florida. Russia is exploring in the Arctic Ocean, near the coast of Alaska.

Obama is calling on us to "protect the environment" by not drilling, while our sworn enemies are using outmoded technology that is guaranteed to create pollution that dwarfs anything American technology could possibly generate.

Obama won't rest until he's brought Big Oil to its knees and crippled our economy in the bargain.

In order to do this, he's not just beseeching other oil-producing nations to ramp up their production, and he's not just attacking the oil companies themselves, and he's not just effectively canceling projects like the Keystone XL pipeline and offshore drilling in American waters.

He's going after commodities "speculators," particularly those that buy and sell oil futures.

It's the third leg of his beg, tax, and persecute strategy to bring the oil industry in the United States under government control.

Tyranny of the Government Regulators

At the center of Obama's strategy of using regulatory agencies to stifle American economic growth is the idea that commodities speculators are the cause of the rise in oil prices. Where energy is concerned, he's relying heavily on two of those agencies, the Commodity Futures Trading Commission (CFTC) and the EPA, to help him implement the government takeover of the energy industry. The two agencies were created as part of the surge of new regulatory bodies in the 1970s, and they represent the rise of regulation as a punitive force in American politics and business.

At the head of these agencies are representatives of two of the cornerstones of regulatory power that form the foundation of Obama's power base: Goldman Sachs and the radical left.

The man he's tapped to go after commodities traders is Gary Gensler.

Here's what you need to know about Gensler, Obama's handpicked choice to run the CFTC.

Like so many others in positions of financial industry power in the Obama administration, CFTC Chairman Gary Gensler is a former Goldman Sachs employee. His job duties involve regulating the multi-trillion-dollar commodities futures market. He oversees the commodities exchanges that trade future contracts for products that range from oil to wheat to derivative financial instruments.

Hiring Gensler to regulate these activities is like hiring the fox to oversee the henhouse.

That's because Gensler, the man now charged with regulating commodities trading, played an important part in creating the financial meltdown that enabled Obama to get elected in the first place. He did that through making sure that close regulation of the trading of financial derivatives did *not* happen prior to 2008.

During the Clinton administration, Gensler was part of the team that saw to it that banks and other financial institutions were able to continue to package subprime mortgage loans and sell them to other financial companies and investors in order to hedge their own positions without close scrutiny. Gensler also helped rescue Long-Term Capital Management, a hedge fund that had gotten in trouble trading financial derivatives.[47] As I explained in *Trickle Up Poverty*, when banks and traders were allowed to engage in selling suspect financial derivatives and in selling securities short with no checks whatsoever, I believe these practices triggered the financial meltdown that resulted in Obama's election.

Did you know that Gensler's policy changes are also responsible in large part for the implosion of Jon Corzine's MF Global? I've told you about Corzine's activities in chapter 4, but you need to know that between 2000 and 2005, Gensler was responsible for loosening the regulations guiding the types of investments futures brokers like MF Global could make with their customers' money. These policy changes resulted in an environment where brokers could put their clients at much-increased risk, and they were instrumental in Corzine's downfall.[48]

Now, Gensler is charged with regulating commodities trading. He's supposed to do exactly what he wouldn't do when he was part of the Clinton team in charge of regulating derivatives trading.

He's the man Obama is depending on to "investigate" the role of commodities traders in the rise of oil prices.

I don't trust him to uncover fraud.

I do trust him to carry out Obama's continuing demonization of the oil and gas industry through attacks on commodities traders. The administration says he's supposed to find them guilty of bilking the public, of making too much money at the public's expense.

In fact, he's supposed to take the spotlight off of Obama's role as the true cause of rising oil prices.

The country is not buying Obama's version of events, though.

Americans understand that Obama's strategy to reduce prices is largely symbolic. The power-grabber in the White House rejects what he calls "quick fix" policies to lower oil prices while insisting that he's focused on the long term.

In a *Washington Post*-ABC News poll, 71 percent of the people responding said they were being hurt by high gas prices, and 57 percent of them found fault with the way Obama was handling the economy. The signal Obama is sending is that "we're not interested in producing oil and gas in this country," according to Thomas Pyle, the president of the Institute for Energy Research.[49]

That's not something Americans want to hear.

What's even more revealing, though, is how little Obama understands of the market forces that are at work in determining oil and gas prices. At the time the president went on his crusade to find fraud among oil speculators, oil supplies were at normal seasonal levels, not an indication that any sort of fraud was being perpetrated. As one industry analyst put it, "This is a transparently political fishing expedition that insinuates that fraud or manipulation is distorting oil prices without providing even the flimsiest factual basis for such a suspicion."[50]

As I've told you, Obama and his henchmen are doing everything they can to keep oil and gas prices as high as possible. That's why Obama is actually banking on Gensler to fail. The subversive in the Oval Office doesn't want to bring down oil prices, and Gensler is just the man to make sure they continue to rise to unprecedented levels.

Rising oil and gas prices are part of their plan to make the United States a second-class nation. One way to do that is to push a "green" agenda, a political strategy that places the interests of environmentalists above concerns about our economic stability and our national security. To accomplish that he's relying on the EPA.

The Environmental Protection Agency was the creation of the Nixon administration in 1970. It was put in place in order to reverse the effects of the pollution spawned by the postwar industrial expansion in the United States, but in the Obama administration, it's been transformed into the governmental arm of the radical environmentalist movement.

Lisa Jackson, the current head of the EPA, likes to brag that she's a "Shell Oil creation." In fact, Shell Oil financed her undergraduate education at Tulane University through a scholarship they provided to her.[51] She's also a radical environmentalist. As the head of the EPA, she promotes "environmental justice." That's based on the idea that U.S. environmental policy disproportionately affects people of color, and we've got to shut down our energy industry so that minorities won't be further harmed by pollution. Jackson's idea of energy policy is to promote government-subsidized "green jobs" so she can create "environmentalists for life." For her, the EPA is synonymous with the environmental movement, and she makes it clear that she will do whatever she needs to in order to implement policies that significantly impede the expansion of our energy industry.[52]

Financing Jackson's education is one mistake Shell Oil would like to take back.

On April 26, 2011, Jackson's EPA denied permission for the Royal Dutch Shell—Shell Oil's parent company—to drill off Alaska's northern coast.

The EPA's reason?

After the company had spent some $4 billion in exploration costs and had gone through the tortuous process of applying for a permit to drill, the EPA found that they had missed something. They hadn't considered the environmental impact of the emissions from an ice-breaking boat they planned to use in the operation. The pollution created by the vessel, the EPA insisted, might contaminate the air and put at risk the inhabitants of a small village about 70 miles away.[53]

That's it.

That's the reason Shell, which had jumped over every other hurdle the EPA put in its way, was denied a permit to drill off the coast of Alaska.

Decisions like this are characteristic of the dictatorial power this administration is wielding over the very industry that is the lifeblood of our energy production, the foundation of our economic well-being.

Obama tapped the EPA to take over this role when the Senate refused, in June 2009, to pass the cap-and-trade legislation that the House had approved in 2008. That legislation had been described as "a centerpiece of President Barack Obama's ambitious effort to transform the way Americans produce and consume energy."[54]

That should read, "President Barack Obama's ambitious effort to intrude further into our lives and wreak more havoc on the U.S. economy through his radical energy agenda."

The radical environmentalists' logic is straightforward: Making the cost of burning coal and oil significantly higher will encourage investment in new energy sources that reduce greenhouse gas emissions. Obama buys into it. He's said as much himself when he explained during the 2008 campaign that under the policies he was proposing energy prices would "necessarily skyrocket."[55]

We should have listened more closely at the time.

We should have gotten the message.

What happened when he couldn't pass his cap-and-tax legislation perfectly defines *trickle down tyranny*: He turned to the Environmental Protection Agency in order to bypass Congress.

He decided to legislate by presidential decree, using the EPA as his enforcer.

The Supreme Court backed him up. Their decision in the 2007 case *Massachusetts v. EPA* ruled that the EPA *does* have the power to regulate greenhouse gases under the Clean Air Act.

The Clean Air Act itself is not quite so clear. For instance, it doesn't recognize carbon dioxide as a pollutant that needs to be reduced. It lists carbon monoxide, sulfur dioxide, particulate matter, and lead among the primary pollutants that need to be cleaned up.[56]

And the United States has done an extraordinary job of removing them from the environment. By the year 2000, carbon monoxide levels were

31 percent lower than they were when the Clean Air Act was passed. Sulfur dioxide (27 percent reduction), particulate matter (71 percent), and lead (98 percent) have all been cleaned up as well.[57]

That hasn't stopped the EPA from implementing regulations that now threaten to cripple the U.S. electric grid altogether.

You remember the freak snowstorm that slammed the East Coast in October 2011. More than 800,000 Connecticut residents were without electric power for nearly a month after that storm!

The reason?

Lisa Jackson and the EPA are doing Obama's bidding in order to shut down the U.S. energy industry.

Buried in the mountains of regulatory documents compiled to make sure Americans don't have enough energy to meet their needs is this: The EPA "is aware . . . that this regulation may detrimentally impact the reliability of the electric grid." The regulation they're talking about is one designed on the face of it to reduce the amount of mercury released into the air as a result of generating electricity. In fact, though, the regulation is designed to shut down coal-fired electricity generation.

In other words, it's designed to reduce the electricity available to Americans in the name of preventing pollution.

It's working. Just ask 830,000 Connecticut residents, or the people who were victims of the rolling blackouts in Arizona and California in September 2011.[58]

In fact, the EPA draft documents contained language that advised the people who run the grid, including regional transmission operators and state regulators, to start planning to construct new power generation facilities, because EPA regulations will soon cause many existing power plants to shut down. This language was edited out of the final draft of the document.

The EPA simply couldn't let the fact that its new regulations threatened the lives and safety of millions of Americans slip out.[59]

It's bad enough that this administration is reducing the electric power available in the name of preventing miniscule amounts of a contaminant to be released into the atmosphere. But it's even worse that it's doing even more harm in the name of preventing a benign gas—CO_2—from being released into the atmosphere.

When are you going to wake up and realize that Obama's energy policy has nothing to do with protecting the environment?

When are you going to understand that Obama's energy policy is nothing more than another part of his multifaceted plan to take down America, to relegate us to second-class status among world powers?

The Myth of Global Warming

I've known for a long time that the foundation on which the idea of man-made global warming has been built is a complete fabrication, based on the phony "science" of corrupt researchers.

I've explained to you that it has absolutely nothing to do with what might happen if we continue to emit CO_2 into the atmosphere and everything to do with the redistribution of wealth from richer nations to poorer.

Contrary to the EPA's claims, CO_2 is *not* a "greenhouse gas." Carbon dioxide makes up only about three-tenths of one percent (0.03) of the earth's atmosphere. It's statistically insignificant in any calculation that might attempt to identify the causes of man-made global warming. Water vapor, on the other hand, is a "greenhouse" gas, but even so its effects on our climate are dwarfed by changes in the energy from the sun received on earth.[60]

One of the problems we face is that we shouldn't be in the position of having to decide "greenhouse gases" need to be regulated in the first place.

That's because, not only do greenhouse gases, especially CO_2, *not* contribute to global warming, there's no such thing as global warming in the first place.

Let me explain to you the real agenda behind this phony policy.

Ottmar Edenhofer is the co-chair of Working Group III of the Intergovernmental Panel on Climate Change (IPCC). He helped organize that group's 2010 UN Climate Change Conference, known as COP-16, held in Cancun, Mexico. In an interview prior to the start of the conference, Edenhofer explained that climate change has nothing at all to do with climate: "[O]ne must say clearly that we redistribute de facto the world's wealth by climate policy. . . . One has to free oneself from the illusion that international climate policy is environmental policy. This has almost nothing to do with environmental policy anymore, with problems such as deforestation or the ozone hole."[61]

But Edenhofer's comments have much more than "economic" implications. What he's really saying is, "Science be damned. We're going to make sure the data support global warming no matter what, so that we can effect the transfer of wealth that we're politically committed to."

When climate science is no longer about the climate but rather about ways to engineer a massive transfer of wealth from developed nations, not to underdeveloped nations but into the pockets of globalists, then it's no longer science at all but a ruse designed to achieve a specified outcome.

You heard me.

Climate science as it is practiced by the left is nothing but a ruse.

It's a scam.

It's designed to make sure that you forfeit any right you have to maintain control over the way you live your life.

Edenhofer's admission that global warming is nothing more than a political issue totally undermines the "science" invented to justify global warming. The sole reason for the theory that man-made global warming exists at all is to help its proponents gain political and financial power over the money generated by capitalist economies.

But beyond that, the "science" itself is a hoax. As I explained in *Trickle Up Poverty*, the Climategate scandal, where it came out that the "data" that supposedly confirmed that there was such a thing as global warming were fabricated and manipulated, finally put a damper on the lies that charlatans like Al Goreleoni were perpetrating. As Ottmar Edenhofer's words demonstrate, the people who perpetrated the hoax have admitted as much.

Global warming is indeed a man-made phenomenon. The men who made it were not the people burning coal, oil, and gasoline, though. They were the so-called scientists who concocted and altered temperature data over the past ten years in order to make it appear as if the rest of us were guilty.

The administration claims that there's a consensus among scientists that global warming is a real and imminent threat to humanity.

That's a lie.

When I hear the word *consensus* I simply point to the 9,000 scientists with legitimate scientific credentials who were among more than 30,000 people who signed a petition that said there was no such thing as man-made global warming.[62]

That's what I call consensus.

But the evidence doesn't stop with scientific "consensus." A project called the CLOUD experiment, in which the European Organization for Nuclear Research (CERN) built a chamber that exactly reproduced the earth's atmosphere, demonstrated conclusively that cosmic rays from the sun—and not man-made activity—are the primary cause of climate change on earth.[63]

Global warming has been debunked as a serious threat to our economic well-being.

What has replaced it is the administration's continuing insistence on making policy decisions based on the idea that we still somehow need to control carbon emissions into the atmosphere.

Despite what Obama would like us to believe, it's his own administration's policies—overseen by leftist haters of capitalism like Lisa Jackson—that are causing the dramatic spike in oil prices, which in turn causes out-of-control increases in the price of gasoline at the pump.

Imposing those policies on the American people follows the leftist playbook. When the fabrication and manipulation of data that was the foundation of the man-made global warming myth was revealed, one of the perpetrators, Phil Jones of the Climate Research Unit in East Anglia, Great Britain, "shaped" the news, lied to the news media, and obstructed requests for the data under the Freedom of Information Act.[64]

Jones was sticking to the plan outlined in a pamphlet titled *The Rules of the Game.* The pamphlet is nothing less than a collection of leftist talking points that were distributed to radical environmentalists about the energy issue. It explains how to misrepresent and obfuscate the issue of climate change in order to avoid public scrutiny.

Included in the "rules" for maintaining the fabrication that global warming is a man-made phenomenon are these:

"Be cool: Be sexy, mainstream, non-patronizing, brave—stand out!"

"Belong: Join a massive worldwide change, start positive conformity, join a success."

"Only stories work: Empathy and emotions are powerful, use stories to hold people's attention."

"Change is for all: Break stereotypes, use inclusive language and images, push mass ownership."[65]

Does this sound like a page from the Marxist playbook?

Something Barack Obama might say in one of his speeches?

That's exactly what it is.

These "rules" are nothing more than the blueprint for furthering an expansion of control over the people by means of perpetuating the hoax of global warming that dwarfs anything in George Orwell's *1984*.[66]

The U.S. Congress is an unindicted co-conspirator in the government's takeover of the energy industry.

Congress v. the EPA

When the Senate decided not to vote on the Waxman-Markey comprehensive climate "cap-and-tax" legislation in 2009, they were aware of the threat that the EPA under Lisa Jackson might go ahead and implement the regulation of greenhouse gases without congressional approval. At the time, the president admitted that the cap-and-tax bill was dead, but at the same time he said that there was "more than one way to skin a cat."[67]

That's exactly what Jackson is preparing to do.

And that's exactly what the Senate has voted to let her do. In April 2011, the House passed the Energy Tax Prevention Act, which would have denied Jackson and the EPA the ability to regulate carbon emissions, by an overwhelming majority, with 19 Democrats going along with every Republican in approving the measure. When it reached the Senate, however, the bill failed to pass. Senators debated on several proposals for limiting the EPA's authority, but none was approved.

Although none of the Senate versions of this important bill was able to generate enough votes to pass, in the voting on all four measures taken separately, a total of 64 senators voted to approve at least one of the amendments. There is a majority in the Senate who want to deny the EPA this overreaching and dangerous authority, but the idiots in that chamber can't find a way to do it.[68]

The failure cleared the way for another Obama growth-killing policy as Congress declined to step in to stop the Obama administration's plan to increase the cost of energy in the U.S.

Obama wants to reduce carbon emissions.

The American people want him to reduce gas prices.

You know whose side he's on.

Now Jackson and the EPA are planning on directly regulating the emission of carbon dioxide and other pollutants from power plants and other industrial facilities, including everything from cement plants to oil refineries.[69]

Green energy is big business.

If you don't believe me, just ask Barack Obama.

But black energy is even bigger business.

Just ask George Soros.

Drill, George, Drill

Don't get me wrong.

George Soros is investing in green energy.

In fact, as part of his green energy investment strategy, he's involved in a move to destroy one of the largest and most important eco-environments in the world: Brazil's Cerrado.

The Cerrado is a "vast plateau where temperatures range from freezing to steaming and bushes and grasslands alternate with forests and the richest variety of flora of all the world's savannas." The area supports some 160,000 species of plants and animals, many of them found only in the Cerrado, a number of them on the endangered list.[70]

Where is Barack Obama when his beloved endangered species are being threatened in Brazil? The same endangered species he and Lisa Jackson use to make sure that U.S. energy companies can't pursue energy strategies in this country that benefit the United States?

He's hiding behind the man who I believe is the architect of his energy policy: George Soros.

Here's how it happened.

When Congress recently mandated that American consumption of ethanol be increased to 36 billion gallons annually in the next decade, the race was on to develop sources for the "alternative" fuel. That meant increasing the production of crops, such as corn and sugarcane, that can be converted to ethanol. That also meant that Soros was, in effect, partnering with former president George W. Bush, who had negotiated a deal with Brazil in 2007 to increase the production of ethanol in South America.[71]

Soros is backing a Brazilian company, Adecoagro, that is one of the primary investors in Brazilian ethanol. The company plans to spend a billion dollars to build three ethanol production plants in Brazil. Soros's company, Pampas Humedas LLC, which is an affiliate of Soros Fund Management LLC, owns about 33 percent of Adecoagro.[72]

That translates to greater destruction of the Cerrado. The worldwide demand for ethanol, fueled by U.S. legislation, has made crops such as corn and sugarcane used to produce ethanol much more lucrative than they were when they were simply food crops. In the process, it has made the production of beef less profitable.

As the need for more corn and sugarcane from which to produce ethanol has skyrocketed like Obama's energy prices, so has the need for land on which to plant them. That means that companies in the ethanol business are clearing and planting tens of thousands of acres in the Cerrado that have previously been home to one of the most diverse ecosystems in the world.

The U.S. controls the ethanol market through tariffs and subsidies. It imposes a tariff of 54 cents a gallon on imported ethanol, while at the same time it subsidizes the production of ethanol in the United States to the tune of 45 cents a gallon. The Senate is now claiming that we can't afford to keep subsidizing ethanol production here. At the same time, Brazil is complaining about the tariffs that keep its products from being able to compete with those produced in the U.S.

California Democratic Senator Dianne Feinstein puts it this way: "It's time we end subsidies that we cannot afford and tariffs that increase gas prices."[73]

José Goldemberg, former Brazilian Secretary of State for Science and Technology, puts it this way: "If the U.S. entirely lifts the tariff, demand for ethanol will go through the roof and the pressure on the environment would be enormous."[74]

I put it this way: If ethanol tariffs and subsidies are removed, Barack Obama and George Soros will have taken another step toward undermining the U.S. oil industry and destroying the environment.

It appears very likely that Obama's energy strategy is being controlled by George Soros. I argued in *Trickle Up Poverty* that I believe Soros was instrumental in putting Obama in office, and Soros is now collecting his fee in the form of energy policy that builds his enormous fortune and increases his power around the world.

You also have to keep in mind that Soros always hedges his bets. He made his fortune as a hedge fund entrepreneur, and he hasn't changed his ways since branching out into the energy industry.

He's hedging his ethanol bets by also investing heavily in oil production. Like Barack Obama, Soros's energy investments are not in the United States.

I've mentioned that during his family junket to Brazil in the early spring of 2011, Obama promised that country's state-controlled oil company, Petrobras (short for Petroleo Brasileiro), $2 billion in American loans to help them jump-start their offshore drilling programs, in the Tupi oil field near Rio de Janeiro.

There's more to the story.

The organization providing financing for the Brazilian offshore oil production to go forward is the U.S. Export-Import Bank. The Ex-Im Bank is part of the U.S. government and so is presumably beholden to American citizens. The idea that Obama would even think about signing off on a move to provide money that puts our own federal bank in direct competition with American oil interests borders on treason.

What makes the move increasingly suspect is that Soros Fund Management LLC, the same company that invested heavily in Brazilian ethanol production, also owns an $811 million stake in Petrobras. That stake in Brazilian oil represents 22 percent of the investments of Soros Fund Management.[75]

Even though, because of new regulatory rules, Soros's company is no longer managing other people's money,[76] Soros continues to invest in energy production for himself and his family.

Soros Fund Management also owns 11.9 percent of an Australian energy company called InterOil. Soros's stake in that company represents the third-largest single investment in his hedge fund.[77]

Now InterOil's primary asset is a vast reserve of natural gas on the island of New Guinea. Development of this gas field is important to the success of Soros's strategy. Soros hopes to curtail production of natural gas in the United States so that we, even though we have the largest recoverable fuel reserves in the world,[78] will have to buy natural gas produced in other countries.

In order to protect his investment and to reduce the competition from U.S. companies, Soros is lobbying vigorously for the U.S. Securities and Exchange Commission to pass rules that would require American oil companies to disclose payments to foreign countries. In practice, the rules Soros favors would make it extremely difficult for American companies to make deals enabling them to drill for gas in foreign countries, and it would put them at a disadvantage against state-backed oil companies in other countries.[79]

Soros also funded a "documentary" screened at ultraliberal Robert Redford's Sundance Film Festival that denounced the developers of shale gas for the impact their drilling had on the environment!

As if Soros, with his investment in the destruction of the Cerrado, has any concern whatsoever for the environment.

Soros openly admits that his primary purpose in making the film was to undermine U.S. production of natural gas.[80]

The combination of Soros's trying to curtail natural gas production in this country and trying to make it more difficult for U.S. companies to produce oil and natural gas in other countries plays into his strategy of crippling the U.S. energy industry.

The perfidious activity doesn't stop there.

The Ex-Im Bank has recently granted yet another loan to yet another foreign country for the purpose of developing oil capabilities that we in the United States desperately need to develop here. The bank, which has become Obama's foreign energy production investment arm, provided $2.84 billion to Reficar, a wholly owned subsidiary of Ecopetrol, the national oil company of Colombia, to enable it to build new refineries. The money involved makes this the second-largest project financed by Ex-Im. The largest is the New Guinea natural gas project. This loan was made despite the fact that, with the exception of a low-capacity refinery opened in Valdez, Alaska, in 1993, there have been no new refineries built in the U.S. since 1976.[81]

Do you know who appoints the board of directors of the U.S.-government-controlled Export-Import Bank?

The president of the United States appoints them.

Do you know who appointed Fred Hochberg, the current president of the board of Ex-Im?

Barack Obama did.

Obama couldn't even wait until he was sworn in to appoint Hochberg. On January 9, 2009, 11 days before his inauguration, Obama made a public announcement of his intention to nominate Hochberg.[82]

The reason Hochberg's appointment was so urgent was that it was part of the payback exacted by George Soros for having chosen Obama as his puppet in the White House.

You see, before his appointment, Hochberg had been Dean of Milano, The New School for Management and Urban Policy, one of the most left-leaning universities in the United States.

The school was formerly known as The New School for Social Research and was founded in 1919 by disaffected leftist professors at Columbia University who objected to the loyalty oaths they were being subjected to in order to prove they weren't communist dissidents.[83] I remember when I was in college during the late 1950s and early '60s just how radical a reputation the New School had. It offered a temporary home for Marxist intellectuals like Herbert Marcuse—"the father of the new left"[84]—when they fled from Nazi Germany.

Its politics haven't changed much in the past half century.

Hochberg, the new head of the Ex-Im Bank, shares their radical leftist views. He also happens to be openly gay, and he's member of the public policy committee of the Human Rights Campaign, an organization championing gay and lesbian rights.[85] He's been touted by the gay journal *Out Magazine* as "the 15th most powerful gay person in America."[86]

George Soros is a frequent visitor to the New School.

In 2007, the school hosted Soros at a panel about how philanthropists spend their money. During that panel, Soros was questioned by Hochberg about the value of contributing to political candidates in the light of Soros's support of John Kerry in the 2004 presidential election. Soros answered that it was "money better spent."[87]

Things have come full circle.

Now Hochberg is the head of a federal bank that has no congressional oversight, and he brings a leftist political philosophy in keeping with that of George Soros and Barack Obama to his job at a financial institution whose choices of where and how much to invest are outside of any regulation, even that of the Constitution.

Its activities are determined by only one person: the president of the United States.

In this case, that means they're determined by George Soros, since the president is effectively Soros's proxy.

The U.S., then, will either print money or borrow it from China and loan it to the Colombians to build refineries that compete with our own. In return, the United States is supposed to receive 15,000 temporary jobs.

Those jobs are projected to last four years.

And you'll never hear of those jobs again after you read this.

The Real Energy Crisis

Obama's energy policies during the first two years of his administration have reflected Soros's vision, a vision developed through his anti-American think tank, Center for American Progress. That organization has been a fountain of leftist, anti-capitalist thinking since it was founded in 2003.

The Obama/Soros energy policy has several components, including promoting green energy, causing oil and gas prices to "skyrocket," stifling oil and natural gas production in the U.S., weakening the dollar, and aiding and abetting the enemy by encouraging and giving international competitors loans through the Export-Import Bank for projects in competition with those promoted by U.S. interests.

Among the most important members of Obama's team of czars was Carol Browner, one of the key people Obama has tapped for his administration as part of his energy strategy. Browner was a member of the Commission for a Sustainable World Society of the group Socialist International, which, as its name indicates, is a leftist political group dedicated to the spread of socialism around the world.[88] Her green energy activities were an extension of her political philosophy, which focuses on the destruction of capitalism.

At the same time, the EPA and the CFTC have been implementing another component of the Obama energy policy: making sure that U.S. oil companies are continually stopped as they try to explore and develop new oil production capabilities. As of February 2011, there were more than 100 oil drilling permits awaiting review by the Bureau of Ocean Energy Management, Regulation and Enforcement. At that time, the Obama administration was issuing an average of fewer than two permits a month.[89]

In addition, Obama has reversed his earlier decision to open U.S. coastal waters to exploration, instead banning any drilling on the Atlantic and Pacific coasts of the United States and in the Gulf of Mexico. The effect of this will be to reduce oil production in the Gulf by more than 200,000 barrels a day at a cost of well over a billion dollars annually in lost revenues as more than 30 Gulf oil rigs shut down, leaving some 6,000 employees out of work. The ancillary effect will be to take away jobs from those who support the oil industry, from restaurant owners to retailers to equipment wholesalers.[90]

The rising gas prices Steven Chu speaks of—remember, it was Chu who said we need to make sure gasoline prices in the United States reach European levels—are an important part of Obama's energy strategy, which is focused on finding ways to make oil and natural gas prohibitively expensive and force us into adopting so-called green energy technology.

Obama is also now floating the idea of a mileage-based tax on gasoline in the form of what it calls the Transportation Opportunities Act. The "opportunities" in the title are for the administration to intrude even further into your life by requiring you to install an electronic device on your car that records the miles you drive, with payment for those miles being deducted from your credit card every time you pay to fuel up at the gas station. The program would be administered by something called the Surface Transportation Revenue Alternatives Office, which would develop the program by doing what the plan calls "a study framework that defines the functionality of a mileage-based user fee system and other systems."[91]

It's a plan that would impose taxes on the poor, increase the cost of gasoline—which is already taxed exorbitantly—and further weaken the oil industry by reducing the number of miles driven by U.S. car owners.

In other words, it's a plan that fits in nicely with the administration's attack on American gas and oil production and sales.

I've explained in chapter 4, "Tyranny of the Treasurer," how Ben Bernanke at the Fed and Timothy Geithner at the Treasury Department have systematically devalued the dollar on world markets. Their policies also have a powerful impact on the price of oil. Simply put, as the value of the dollar goes down, the price of the oil we have to purchase on the international market goes up. U.S. financial policy works hand in hand with the Obama/Soros energy policy to drive the price of gasoline up, with no end in sight.

You can't mention the name Barack Obama without mentioning the name George Soros at the same time. The two are the political equivalent of Siamese twins. They're ideologically joined.

The list of Ex-Im-funded energy projects is a perfect representation of the Obama-Soros anti-American energy strategy. It focuses on production in several key energy industry segments: corn/sugarcane for ethanol production, oil, and natural gas. What stands out is that this strategy focuses on those industries in every area of the world but the most important one: the United States.

I maintain that George Soros is dictating energy policy in many key areas, and his confederate in the White House is putting his vision into practice.

Obama and Soros are expanding their anti-American investments by making sure that their appointee at Ex-Im, Fred Hochberg, steers investments funded by a U.S. federal bank outside the United States to companies and countries that are openly competing with us in the energy market.

Of course, Obama's obeisance to George Soros is a key factor in this plot. Soros is a self-described enemy of America. In fact, he's said, "The main obstacle to a stable and just world order is the United States."[92]

What Soros means by a "stable and just world order," though, is a global government. Unfortunately, I believe that our president, along with Soros, wants nothing less than to see the demise of the greatest country in the world as he sets his sights on being part of a global government built on planetary crony capitalism.

It's all part of the globalist agenda that informs Obama's political vision:

He's promoting the demise of the U.S. as a world power in order to bring into being a global government, run by a handful of powerful autocrats like George Soros, in league with crony capitalist corporations. Their vision is for a world without borders, a world without the U.S. standing in the way of their grabbing and maintaining power over the lives of every citizen on the planet.

CHAPTER 10

Tyranny of the Anti-Justice Department

On October 12, 2011, Eric Holder announced that the United States had uncovered a plot to assassinate the Saudi Arabian ambassador to the United States, Adel al-Jubeir—on American soil! The plot was supposedly hatched by the Iranian Quds secret service with the knowledge of President Mahmoud Ahmadinejad. Although one of the suspects, Manssor Arbabsiar, was arrested in September, Holder didn't make the announcement until the next month, only days before he was to be subpoenaed to appear before a congressional committee to testify in the Fast and Furious guns-to-Mexico scandal that had come to light in the previous six months.

Holder's report about the assassination attempt included this extraordinary information: The Iranian secret service was said to be trying to hire the Mexican Zeta drug cartel to perform the assassination.

Apparently they would be using weapons that Holder's Bureau of Alcohol, Tobacco, Firearms & Explosives (ATF) sold to them as part of the Fast and Furious guns to Mexico operation in order to assassinate the Saudi ambassador.

That would complete the cycle of corruption.

Holder, through the ATF, would sell weapons to Mexican drug cartels. This would help the cartels increase their presence and their firepower, making it easier for them to transport and sell drugs in the United States.

This would in turn undermine federal agents and state drug enforce-ment agencies' abilities to fight the war against drugs. It would also exacer-bate our illegal immigration problems, which Holder obviously wanted to see happen—he'd fought state laws such as Arizona's that tried to enforce existing federal laws against illegal immigration.

And finally, a trumped-up assassination charge involving weapons that Holder had ordered "walked" over the U.S. border into Mexico and sold to drug cartel members would allow Holder to divert attention from his in-volvement in the guns-to-Mexico scandal.

Does that sound like something a patriot—someone who has the best interests of the United States at heart—would do?

Consider this: Under Holder, criminals like illegal aliens and black sepa-ratists from organizations like the New Black Panther Party are no longer criminals. The criminals are now the people like Eric Holder who enforce what's left of the law they're undermining.

I find it hard to believe that there was some sort of plot afoot by Iran to assassinate the Saudi ambassador, not that Iran wouldn't mind seeing him dead.

I do believe the plot could have been hatched in the Attorney General's office in order to divert attention from his appearance in front of a congres-sional committee. He had a difficult job: He had to explain how he—the Attorney General of the United States—could not have known about a gun-running operation—Project Fast and Furious—that had been announced publicly by one of his own staff nearly two years before.

Iran's response to the accusation was probably closer to the truth than anything Holder's Justice Department said in their presser. A spokesman for Mahmoud Ahmadinejad said, "They want to take the public's mind off the serious domestic problems they're facing these days and scare them with fabricated problems outside the country." [1]

In the Fast and Furious scandal, members of Eric Holder's Justice Department—including Holder himself—clearly have lied repeatedly under oath as they try to cover up the illegal sales and transportation of guns to Mexico. These guns were used to kill at least two Americans and hundreds of Mexican citizens.

Why would the Holder Justice Department sell guns to Mexican drug cartels and then intentionally not track where the guns ended up? It was part of a program designed to enable Obama to crack down on gun sales in the U.S.

I'll give you more details later in the chapter, after I make you aware that Fast and Furious is not the only scandal Holder's DOJ has been involved in.

Who Is Eric Holder Working For?

I explained to you that the crash of the Chinook helicopter in more than 20 Navy SEALs were killed very probably happened because the flight was sabotaged by someone in the Obama administration.

I've also told you that Barack Obama and Eric Holder have not instituted an investigation of that incident.

Instead, they're focusing on investigating our own patriots.

Did you know that our Attorney General is actively prosecuting two CIA interrogators who extracted information that helped our forces locate and take out bin Laden?

First this administration manages to get more than 20 Navy SEALs killed in a helicopter crash, then they prosecute the CIA agents who helped the SEALs do their job.

Holder is pursuing these agents despite the fact that they were exonerated in two earlier Justice Department investigations. The CIA agents Holder is singling out for prosecution and others like them have saved countless lives around the world. Intelligence gathered by these interrogators stopped dozens of terrorist attacks, including a plot to blow up the U.S. Embassy in Paris, a plot to sink U.S. warships off the Straits of Gibraltar, and another to blow up stadiums during the World Cup soccer tournament.[2]

Barack Obama paved the way for the criminal prosecution of the two agents when he issued an executive order that said federal law enforcement agents *are bound by the Geneva Convention* in the war against Islamist terrorists, even though the terrorists they're dealing with are not.[3] This order reversed a 2002 Justice Department ruling and freed Holder to prosecute the agents. Now the CIA agents, defenders of America, will be forced to use their personal savings and forfeit their homes to pay the huge legal fees they'll no doubt pile up defending themselves against their own Justice Department. Holder is turning on the very people who saved American lives from al Qaeda terrorist attacks after 9/11.

We have to ask is this: Which side is Holder is on?

It doesn't stop there. Holder has been trying to undermine the United States in the war against Islamist terrorists for a long time.

Did you know that when he worked at the law firm of Covington & Burling prior to becoming attorney general, Holder himself represented al Qaeda terrorists in court—at no charge to the terrorists?

Holder also appointed nine attorneys to posts in his Anti-Justice Department who had provided free legal services for terrorists.[4] One of his new hires, Principal Deputy Solicitor General Neal Katyal, was instrumental in getting the Supreme Court to throw out the military trial system that had been established under George W. Bush. That meant that the U.S. had no venue for trying terrorists until Congress voted to reauthorize the system. And it was Katyal who insisted in his defense of al Qaeda terrorists that they should be treated the same as legal aliens who hold green cards.[5]

Now Holder is being accused of conflict of interest, because he's supposed to prosecute the very terrorists he defended in court.

Shouldn't we be accusing him of conflict of allegiance instead?

Barack Obama has teamed up with Holder to help out the terrorists by putting the Attorney General, a civilian, in charge of overseeing the military trial of 9/11 mastermind Khalid Sheikh Mohammed.

Did you know that appointing a civilian to supervise a military court process is unconstitutional? Obama did it to stall the trials of the 9/11 conspirators until after the 2012 presidential elections. The president had to lie in order to throw America off the track. He knew Americans were worried that he wouldn't prosecute the 9/11 terrorists, so shortly before he was inaugurated Obama met personally with several 9/11 victims' families and told them that he would not stop the military trials of Khalid Sheikh Mohammed and four others. Two days after he was sworn in, Obama went against his word to those family members, ordering Holder to try the terrorists in civilian court—even though their military trials were already scheduled![6]

Did Obama put Holder in charge of the military tribunals in order to *stop the prosecution of the terrorists at Gitmo*?

Has a civilian ever been placed in charge of a military process such as this?

Is Holder going to pardon the Gitmo terrorists like he pardoned convicted Clinton fund-raiser Marc Rich, and then release them with compensation because they were freedom fighters and we illegally detained them?

Is this part of Obama's overall plan to grant amnesty to illegal aliens in order to get reelected?

Holder's DOJ has coddled the people who finance terrorism around the world, too.

The Islamic Investment Company of the Gulf (IICG) is a wholly owned subsidiary of Dar al-Maal al-Islami Trust, an Islamic financial company with two Muslim Brotherhood board members. The company has ties with another Islamic bank that has been investigated for links with terrorist financiers. Holder's DOJ should have been aggressively pursuing criminal action against the group. Instead, he has requested a stay of discovery to prevent information that is vital to other civil cases against IICG from being revealed. It's not clear whether Holder's DOJ ever reached a settlement with IICG, or whether they just dropped charges altogether, as it did against the New Black Panthers. Holder refuses to say, despite requests from a House Appropriations subcommittee to release the records.[7]

Is Eric Holder covering up the fact that he dropped a criminal case against IICG because he's in sympathy with the aims of Islamist terrorists and those who sponsor them?

Holder is supporting the forces working against American interests in other ways, too.

Did you know that he dropped charges against former DOJ lawyer Thomas Tamm, who had been indicted by a federal grand jury for leaking classified information to *The New York Times* about a Bush administration wiretapping program designed to intercept terrorist calls between high-level al Qaeda operatives in Pakistan and their contacts in the United States?

The *Times* won the Pulitzer Prize for the piece, and America's vital wiretap program was shut down. Thanks to Holder, Tamm got off scot-free.

Compare this to what's happening with Rupert Murdoch. As I told you in chapter 3, "Tyranny of the Government-Media Complex," Murdoch is being prosecuted by the Justice Department for alleged wiretapping by one of his newspapers. Yet the *New York Times* does something far worse than anything Murdoch thought of doing—they released classified information to the public that compromised our national security by destroying a critical program designed to gather intelligence about our terrorist enemy—and the Justice Department refuses to prosecute?

This is something you'd expect to see in countries like Fidel Castro's Cuba or Hugo Chavez's Venezuela, but it's going on right here, under our noses. We have real tyranny at Eric Holder's Anti-Justice Department, and it's becoming worse by the day.

Now do you see why I'm asking which side Eric Holder is working for?

When I take you through what Holder's Anti-Justice Department has done in the "guns to Mexico" scandal, you'll understand how serious these questions are.

Project Fast and Furious

In a press conference on *March 24, 2009*, Deputy Attorney General David Ogden announced that "the president has directed us to take action to fight these [Mexican drug] cartels and Attorney General Holder and I are taking new and aggressive steps as part of the administration's comprehensive plan. . . . ATF is . . . adding 37 new employees and ten new offices *using $10 million in Recovery Act funds* . . . to fortify its Project Gunrunner, which is aimed at disrupting arms trafficking between the United States and Mexico." Ogden also announced that ATF would be working with Mexican authorities to implement this expansion of Project Gunrunner.[8] Less than a month later, Holder gave a speech in Cuernavaca, Mexico, in which he boasted to his audience about the launch of the Project Gunrunner expansion.[9] Obama talked about expanded "gun tracing" and "gun enforcement policies" in a joint press conference with Mexico's president, Felipe Calderon, in the same month.[10]

Did you get that?

The White House committed *taxpayer money in the form of stimulus funds*[11] for what would become Project Fast and Furious in March, 2009, less than two months after Obama was sworn in, and Holder's Anti-Justice Department was to oversee the operation.

Eric Holder and Barack Obama are still insisting they knew nothing about Fast and Furious.

Within a few months after the launch of Fast and Furious, reports began to surface about U.S. firearms—including AK-47 and AR-15 assault weapons and Barrett .50-caliber sniper rifles—being recovered in large numbers from violent crime scenes in Mexico. Agents at the ATF, which is under the direct supervision of Holder's Justice Department, alerted the office of the U.S. attaché to Mexico about what was happening, but senior leadership at ATF chose not to respond directly to the reports, informing their field agents only that everything was "under control." [12]

When ATF field agents confronted their supervisors about the rising violence in Mexico resulting from the increase in U.S. guns going into the country, the bosses replied that field agents who didn't go along with the program were "in the wrong line of work." [13]

ATF leadership downplayed these reports because the leaders themselves were deeply involved in what had by then become the covert operation known as Project Fast and Furious, which Holder's DOJ and the ATF had put into action at the direction of Barack Obama. The new operation didn't just supplement Project Gunrunner, it transformed it from a program designed to monitor the illegal sales of weapons along our southern borders into one that put those weapons directly into the hands of the drug cartels.

For Project Fast and Furious, a special ATF strike force called Group VII was directed to monitor Arizona gun shops, capturing surveillance video of illegal sales of guns to "straw buyers," proxies who bought the weapons for their eventual drug cartel owners. At least two of the straw buyers were known to be convicted felons, yet they were encouraged by the ATF to buy the weapons. The weapons would then be traced as they were "walked" across the border to Mexico and delivered to drug cartels.

By late 2010, more than 500 Mexican citizens had been killed with U.S. weapons sold to straw buyers and walked across our southern border. Despite Deputy Attorney General Ogden's initial insistence that Mexican authorities would be involved in the new operation, Project Fast and Furious was carried out without their knowledge, and they were outraged about the deaths of their people at the hands of drug cartel members using weapons Holder's and Obama's policies had placed in their hands.[14]

Operation Fast and Furious transformed Project Gunrunner into a chaotic, agenda-driven political fiasco that abandoned its original mandate and became another egghead nightmare put together—and then covered up—by the president in collusion with Holder's corrupt and incompetent Anti-Justice Department.

Holder ordered one of his own Inspector Generals to review Project Gunrunner, but the IG's report made no mention of Project Fast and Furious. There are two possible explanations as to why the administration staged this initial cover-up. It was either intentional, because the IG was told by Holder not to reveal it, or unintentional, because it had been covered up so effectively by Holder's lieutenants that word of it never reached the Attorney General.

I believe the first explanation.

The reason Holder ordered a cover-up was that Project Fast and Furious had turned into the very thing it was supposed to eliminate: a firearms trafficking ring.

The U.S. government was involved in trafficking guns to Mexican drug cartels.

I find it to be one of the most blatantly subversive criminal operations ever engaged in by any administration in the history of this country.

The stated purpose of the joint White House-Justice Department-ATF Project Fast and Furious was to trace the gun traffic in order to stop it by apprehending the drug cartel kingpins to whom the guns were delivered.[15] In the process, though, DOJ and ATF higher-ups instructed Fast and Furious operatives to continue surveillance but to *cease interdiction*.[16]

The real reason for ATF allowing high-powered weapons to be walked to drug cartel kingpins in Mexico is far more disturbing: It was designed to generate data so that Barack Obama could "prove" that U.S. gun shops were responsible for illegal weapons sales and needed to be shut down.

The Obama administration had, for a long time, falsely asserted that "more than 90 percent of the guns recovered in Mexico come from the United States, many from gun shops that line our shared border."[17] If great numbers of U.S. weapons showed up at crime scenes in Mexico, the administration's anti-gun position would be bolstered. Some 2,000 high-caliber assault weapons[18]—some sources report up to 3,000[19]—were ultimately sold and delivered in this manner.

The administration's intent was to use Project Fast and Furious to confirm their bias against the Second Amendment, to demonstrate that gun ownership is dangerous and that the gun shops in Arizona needed to be shut down. That would be the first step, as I think the administration's logic concluded, to their being able to argue for the closing of gun shops all around the country.

One ATF agent spoke out to his superiors soon after Project Fast and Furious began, saying he "had no question that the individuals we were watching were acting as straw purchasers and that the weapons they purchased would soon be trafficked to Mexico and/or other locales along the southwest border or elsewhere in the United States, and ultimately these firearms would be used in a violent crime."[20] The agent asked his supervisors "if they were prepared to attend the funeral of a slain agent or officer after he or she was killed with one of those straw purchase firearms." The supervisors showed no concern.[21]

It wasn't long before his question was answered with the death of U.S. Border Patrol agent Brian Terry.

Terry was described by others who knew him as a "cop's cop." A former U.S. Marine, he was a member of an exclusive fraternity, a BORTAC (Border Patrol Tactical) unit, the Border Patrol's equivalent of a police SWAT team. BORTAC is

the global special response team for the Department of Homeland Security's (DHS) Bureau of Customs and Border Protection (CBP). Its mission is to respond to terrorist threats of all types anywhere in the world in order to protect our nation's homeland. Its agents are counted among the nation's most dedicated and highly trained special operators. Since its inception in 1984, BORTAC has developed a reputation in the special operations community as one of the premier tactical units in law enforcement.[22]

Late in 2010, while on patrol near Nogales, Arizona, Terry's BORTAC group encountered a "rip crew." A rip crew is a group of illegal alien border bandits who "prey on smugglers and undocumented immigrants entering the United States."[23] Two of the rip crew members were armed with assault rifles with the express intent of confronting and killing U.S. border agents. The Mexican rip crew had themselves been "patrolling" the area *inside the*

United States, lying in wait for potential victims, when they encountered the group of BORTAC agents of which Terry was a member.[24]

On the night of December 14, 2010, one of the agents in Terry's group, using thermal night-vision binoculars, spotted this small group of men. He identified himself and ordered the illegals to drop their weapons. Here's what happened when the illegals refused to comply: "[T]wo Border Patrol agents deployed 'less than lethal' beanbags at the suspected aliens. At this time, at least one of the suspected aliens fired at the Border Patrol agents. Two Border Patrol agents returned fire, one with his long gun and one with his pistol. Border Patrol agent Brian Terry was shot with one bullet and died shortly after. One of the suspected illegal aliens, later identified as Manuel Osorio-Arellanes, was also shot."[25]

Beanbags?

Did you get that?

With the takeover of the Obama regime in 2009, the president began installing his confederates in positions of power. As part of his seizure of the U.S. government, he put BORTAC under the jurisdiction of new Homeland Security Secretary Janet Napolitano. In order to make it more difficult for Border Patrol agents to defend themselves and to give an advantage to Mexican drug cartels, one of Napolitano's lieutenants issued a standing order that BORTAC team members were "to always use ('non-lethal') beanbag rounds first before using live ammunition."[26]

Agent Terry and his fellow BORTAC patrol members were required to first fire blanks at the bandits they encountered before they fired live ammunition. The BORTAC patrol's firing beanbags alerted the smugglers to the agents' location.

Terry was killed before he could fire a live round at the illegals.

The serial numbers on the two assault rifles the ATF initially reported found at the scene of Terry's murder were identical to two rifles ATF watched a straw buyer named Jaime Avila purchase in a Phoenix gun store.[27] It wasn't until secret audio recordings made by the lawyer of a Glendale, Arizona, gun shop owner who believed federal agents were lying to his client were obtained by one news organization that we found out there were three—and not two—Fast and Furious weapons found at the scene of Brian Terry's murder.[28] The weapon used to kill Terry was one of the thousands that had been transported across the border to Mexico as part of Project Fast and Furious.

Do you remember how Holder's Anti-Justice Department responded to the murder of one of its ATF agents?

They arrested three of the four suspects, then let them go free.

The only reason they didn't release all four was that one of them was wounded and still in the hospital.

A month after Terry was murdered, three of the four suspects rounded up near the scene immediately after the shootout were indicted by the U.S. attorney's office, but only on charges that they were in the country illegally, and not for complicity in Terry's murder. In February, two months after Terry's murder, the suspects pled guilty to misdemeanor charges and were released for deportation.[29]

It took the DOJ *five months* before they even issued an indictment against Osorio-Arellanes, the one who had been hospitalized since the incident because of the wounds he suffered. Two other unnamed suspects, who were not captured and remain at large, were charged *in absentia* with second-degree murder and conspiracy in connection with Terry's murder.

In the five months before they reluctantly issued indictments in the case, Holder and the ATF did everything in their power to cover up their role in the agent's death. Between the time of Terry's murder and mid-April 2011, California Congressman Darrell Issa and Iowa Senator Charles Grassley sent dozens of letters either directly to Holder or to ATF and other administration officials requesting documentation relating to Project Fast and Furious. Issa complained that the "unwillingness of this Administration—most specifically the Bureau of Alcohol, Tobacco, Firearms, and Explosives—to answer questions about this deadly serious matter is deeply troubling."[30]

When he wasn't stonewalling Issa and Grassley, Holder was orchestrating a campaign of lies about Terry's murder.

Lie #1: On February 4, 2011, Holder's Assistant Attorney General Ronald Weich sent a letter to Grassley saying the idea that "ATF 'sanctioned' or otherwise knowingly allowed the sale of assault weapons to a straw purchaser who then transported them into Mexico is false. ATF makes every effort to interdict weapons that have been purchased illegally and prevent their transportation into Mexico."[31]

Lie #2: In early March, the U.S. Embassy posted a bulletin on its website that mischaracterized Mexico's knowledge of the project, going so far as to state that representatives of the Mexican government were "present" when an arrest of 19 people involved in one sting was made. The Mexican government was quick to respond, saying "[t]he Mexican government has not given, and will never give its tacit or explicit approval for the deliberate flow of weapons into Mexican territory." [32]

Lie #3: On May 3, when he testified in front of Issa's House Judiciary Committee, Holder himself committed perjury. Despite the fact that Holder's Deputy Attorney General David Ogden had announced in March, 2009, that Holder's DOJ was involved in the expansion of Project Gunrunner and that it had been exchanging letters with Issa and Grassley on the subject of Gunrunner and Operation Fast and Furious since the beginning of 2011, Holder claimed he had found out about Fast and Furious "only in recent weeks."

As one blogger said about the Attorney General's blatantly false testimony:

> Holder denied knowing anything about the project, even though records show it started in 2009. He also claimed that he didn't know who had given authorization for the operation. . . . Given Holder's testimony, he wants us to believe that he had no knowledge of an operation that his assistant approved a wiretap for over a year ago, and he only learned of it within the last three weeks? Come on. Either Holder is covering up or he's been too busy SUING Arizona over its illegal immigration laws to see the massive FUBAR his department has created, you take your pick. [33]

It wasn't until ATF field agents themselves began to go public that the truth started to come out. In March, 2011, ATF agent John Dodson, who had been removed from Group VII in the summer of 2010 because he was so vociferous in his complaints about Operation Fast and Furious, appeared on a CBS television show. When asked if he intentionally allowed guns to be walked into Mexico, Dodson simply said, "Yes ma'am. The Agency was." [34] He explained further, "I'm boots on the ground here in Phoenix.

We've been doing it [letting guns be walked across the border] every day since I've been here. . . . Tell me I didn't do that."

When asked by an interviewer if he saw the relationship of the increased weapons flowing into Mexico as a result of Project Fast and Furious, Dodson responded, "I even asked them if they could see the correlation between the two: The more our guys buy, the more violence we see down there." [35]

The straw buyers were aggressive in their purchases. One of them, Uriel Patino, who supplemented his income as a buyer of illegal weapons with food stamps at the expense of American taxpayers, personally bought at least 316 weapons, with some investigators claiming the number could be as high as 600 or more, since ATF accounting is suspect. The weapons purchased by Patino included AK-47 assault rifles—which he often purchased 15 to 20 or more at a time—and dozens of FN 5.7 mm pistols. [36]

Issa summed up the effect of the agents' testimony this way: "ATF agents have shared chilling accounts of being ordered to stand down as criminals in Arizona walked away with guns headed for Mexican drug cartels. With the clinical precision of a lab experiment, the Justice Department kept records of weapons they let walk and the crime scenes where they next appeared. To agents' shock, preventing loss of life was not the primary concern." [37]

Throughout the process of covering up Project Fast and Furious, Holder's DOJ suppressed evidence, intimidated witnesses, and withheld information. Incredibly, by September 2011—six months after word of Project Fast and Furious began leaking out—Holder's DOJ had not produced a written report. The FBI, the Arizona U.S. attorney's office, and Janet Napolitano's Homeland Security Department hadn't put a single word in writing about the project, either.

A recent DOJ letter that Assistant Attorney General Robert Welch sent to Senate Judiciary Committee leaders Patrick Leahy and Charles Grassley revealed that Fast and Furious weapons had been found at a total of 11 crime scenes in the United States. [38] By September 2011, the number of U.S. crime scenes where Fast and Furious guns were recovered was up to 21. [39]

In order to avoid attempted intimidation by Holder's DOJ, acting ATF director Kenneth Melson appeared secretly on July 4, 2011, before a congressional committee convened by Issa and Grassley—weeks before his scheduled testimony—with his personal attorney present, and provided damning information about Project Fast and Furious and Holder's and Obama's involvement. Melson testified that he had reviewed "hundreds of documents . . . including wiretap applications and Reports of Investigations"; until then, he hadn't known the full story, which made him "sick to his stomach." He reported to the Office of the Deputy Attorney General that a review and reexamination of how the DOJ was responding to requests for information by Congress needed to be done.

According to Melson, the DOJ had directed him not to respond to Congress but instead "took full control of replying to briefing and document requests from Congress," cutting ATF out of the loop as it sent false denials to Congress, effectively obstructing the investigation. The information concealed by the DOJ included the involvement of other agencies such as the DEA and the FBI, both of which are under the supervision of the Department of Justice.

The essence of Melson's testimony was communicated to Holder in a July 5 letter Grassley and Issa sent to the attorney general. Among other things, they accused Holder of seeking "to limit and control his [Melson's] communications" even though "direct communications with Congress are so important and are protected by law." [40]

The final Issa/Grassley committee report issued on July 26, 2011, described Operation Fast and Furious this way: "Operation Fast and Furious made unprecedented use of a dangerous investigative technique known as 'gunwalking.' Rather than intervene and seize the illegally purchased firearms, ATF's Phoenix Field Division allowed known straw purchasers to walk away with the guns, over and over again. As a result, the weapons were transferred to criminals and Mexican Drug Cartels." [41]

While all this was going on, Holder's DOJ continued to refuse to release the documents requested by the Issa/Grassley committee.

By the end of August, Issa finally said what had been obvious for months: "We know we are being gamed and we think the time for the game should be up." He explained that his House Oversight Committee had confirmed that administration personnel knew that Fast and Furious weapons were going directly to drug cartels and not being tracked. Many of those involved were "high up in Eric Holder's office and had a lot to do with this [gunwalking] happening." [42]

A week later, Holder was still hanging on to the lie that he knew nothing about Operation Fast and Furious as it was being carried out. As usual he was evasive when he spoke in his own defense: "The notion that somehow or other this thing reaches into the upper levels of the Justice Department is something that . . . I don't think is supported by the facts. . . . I don't think that is going to be shown to be the case."

He doesn't *think* it's supported by the facts? He doesn't think it will *be shown to be the case*?

You notice that he didn't deny anything. He only said that he was pretty sure Issa couldn't pin it on him.

Holder's Continuing Assault on America

Michael Levine, a 40-year Border Patrol veteran, is adamant that "this bizarre idea [Project Fast and Furious] should have [been] killed at birth," wondering if it even included "a way of fully accounting for the number of deaths it might cause." Levine explains that he "would have recommended the immediate dismissal of anyone who either propagated or approved the plan as unfit to be in law enforcement." He would also have "recommended that all those responsible for this plan be prosecuted for the murders of all those identified as having been killed with one of the F&F guns, including but not limited to the murder of Border Patrol Agent Brian Terry." [43]

As another commentator put it, "[W]ith the law enforcement resources responsible for that tracking having been kept ignorant of the entire operation, the alleged rationale for F&F falls flat on its face. [I]n the present situation, we have illegal arms-smuggling, apparently to further a domestic political gun control agenda, which . . . resulted in the deaths of Mexican citizens and two United States federal officers. Now that is a crime, a real and true crime, and one for which Eric Holder and Barack Obama should pay." [44]

We've got a president who funded the operation, then bragged about it in a joint press conference shortly after, and an attorney general who put the operation together and whose Anti-Justice Department managed it for almost a year and a half. After it resulted in the absolutely predictable death of one of our own Border Patrol agents, both the White House and the DOJ denied they knew anything about it, stonewalling federal legislators who tried to uncover the truth. Dozens of people have come forward to testify against Holder and Obama, but they've effectively silenced any attempts to hold them accountable.

Do you need further proof that Obama and Holder were involved in Project Fast and Furious or that the operation was part of a larger plan to expand federal gun control?

Holder's Anti-Justice Department has done exactly that. The DOJ issued new regulations on gun shop owners along the southern border, warning them to report the very activity—the sale of multiple weapons—that the DOJ directed ATF to engage in: intentionally allowing multiple weapons to be sold and then transported to drug cartels in Mexico while their supervisors ignored or punished agents who tried to blow the whistle on the illegal activity. Deputy Attorney General James Cole issued a statement explaining that the new regulations are necessary because "[t]he international expansion and increased violence of transnational criminal networks pose a significant threat to the United States."[45]

Mission accomplished.

Gun control regulation expanded.

Obama and Holder and their ATF confederates officially decreed new gun regulations based on data compiled from illegal activities they condoned and which cost the lives of hundreds of Mexican citizens and two U.S. agents.

Not a single Mexican drug cartel kingpin was arrested as a result of Project Fast and Furious.

What did result from Fast and Furious was a dramatic degradation of the security of our border with Mexico, as drug criminals solidified their hold on the border and extended their authority into U.S. territory.

And what about the key ATF players who worked with Holder and Obama to hide the gunwalking and make this new regulation possible?

For making sure that Obama and Holder were able to implement their sought-after regulations to report multiple sales of long guns in border states, three supervisors—William Newell, William McMahon, and David Voth—were given promotions, despite the fact that I believe they engaged in criminal activities themselves.[46]

Ken Melson, who blew the whistle on Holder and the Justice Department's involvement in Project Fast and Furious, was given a make-work job as "senior adviser on forensic science in the [Justice] department's Office of Legal Policy."

A key resignation tendered so far in the Fast and Furious scandal is that of U.S. Attorney Dennis Burke. His final act before he stepped down? Burke opposed the request of Brian Terry's family when they asked to qualify as crime victims in the case against the gunwalker who purchased the weapons used in their son's murder.[47]

Is this another step toward the ultimate goal of the Obama administration to eliminate the border between the United States and Mexico?

In *Trickle Up Poverty* I wrote about Obama's relationship with the radical Latino group La Raza, with which the president shares the idea that our southern border should be eliminated. Here's what the president said in a July, 2011, address to the group: He confessed that he'd like to "bypass Congress and change the [immigration] laws." He added, "Believe me, the idea of doing things on my own is very tempting. I promise you."[48]

The president forgot to mention that he and his Attorney General were already "doing things on their own" where our southern border is concerned.

Project Fast and Furious was very likely part of Obama's larger "vision" of illegal immigration. In a campaign fund-raiser in San Francisco, the president said this: "No matter who you are. No matter where you came from. No matter what you look like. No matter whether your ancestors landed here on Ellis Island or came here on slave ships or came across the Rio Grande, we are all connected. We will rise and fall together. That's the vision of America I've got, that's the idea of the heart of America. That's the idea of the heart of our campaign."[49]

Do you see what the president is saying when he gets up in front of an audience and claims those coming across the border illegally are the same as those who came to this country through Ellis Island?

Is our president unaware of the law? Or is he just flouting it?

Are you aware that the president's leftist Labor Secretary, Hilda Solis, has signed agreements with a group of Latin American nations that protect the rights of illegal immigrant laborers in the United States? The agreements help immigrants, whether or not they're here illegally, take advantage of the U.S. school system to receive free educations. They also define "abuse" as not being paid the same wages as legal workers and as being deported if they're found out.[50]

And did you know that Obama appointed Cecilia Munoz, a former Vice President of La Raza, as White House Director of Intergovernmental Affairs. I wrote extensively about this organization in *Trickle Up Poverty*. It's the radical Hispanic advocacy group that promotes the elimination of our southern borders, the takeover of several states on our Mexican border by Hispanics, and amnesty for illegal aliens. That's the person Obama appointed to coordinate and oversee the process of developing and executing domestic policy.[51]

The president is doing nothing less than advocating that we give up control of our borders. He's encouraging criminal activity.

It's not working out very well for American citizens. Our southern border has fallen into the hands of the enemy.

Did you know that "birth tourism" is a fast-growing industry in the United States?

You know what birth tourism is, right? That's when an illegal alien enters the country without permission in order to give birth to the baby she's carrying. The kid automatically becomes an American citizen in that case, and the rest of the family can come here "legally."

Did you know that a growing industry caters to these people?

There's a growing number of birth centers, often in upscale suburbs, that charge women tens of thousands of dollars to come here to give birth to their babies so they and their families can become citizens automatically. One owner of a birth center in Queens, New York, said she "spoke to a lawyer before opening my business to make sure I would not get into trouble with U.S. authorities."[52]

These people are taking advantage of *jus soli* laws in order to gain U.S. citizenship through the back door by taking advantage of the Fourteenth Amendment. That amendment was added to the Constitution after the Civil War, and it was designed to make sure that the children of slaves— who had been freed but who were not citizens themselves—would be given U.S. citizenship.

The language of the amendment is open to debate. It says, "All persons born or naturalized in the United States, and subject to the jurisdiction thereof, are citizens of the United States." Visitors to the United States, whether they're here legally on visas or illegally, are not fully subject to all U.S. laws.

It's time the abuse of the Fourteenth Amendment was challenged in court and the fiasco of birth tourism ended once and for all. The law should apply only to children born to legal residents of the United States, and only those here legally and permanently should be covered under the law.

It won't happen under a president determined to give our southwestern states to Mexico.

Homeland Security Secretary Janet Napolitano has announced that her department will only focus on deporting illegal aliens who have committed crimes or who pose a threat to national security or public safety. In the tradition of the Obama administration, she's legislating by decree, and this decree means that being in this country illegally is not a crime.[53]

At the same time, the Department of Health and Human Services announced it would use Obamacare funds to award $28.8 million in grants to 67 community health centers in order to provide free health care to migrant workers. The organizations receiving the grants will not check the immigration status of people who use their services.[54]

Holder's Justice Department is involved up to its eyeballs in the attempt to eliminate borders and encourage illegals to come to America—risk-free. The DOJ was already suing Arizona and Alabama to prevent those states from implementing their strong anti-illegal-immigration laws. In November 2011, Holder and Co. filed a lawsuit against Utah, charging that the law usurps federal authority—authority Holder and Napolitano have repeatedly refused to exercise.[55] At the same time, Holder is prosecuting a U.S. Border Patrol agent for apprehending a drug smuggler with hundreds of pounds of marijuana. The Justice Department managed to convict and send to prison Jesus E. Diaz Jr. on the testimony of the smuggler himself, who was set free.[56]

While all this is going on, illegal immigrants are bilking the federal government of billions of dollars for "earned income tax credits" and "additional child tax credits."[57]

Are you starting to get the picture?

Are you beginning to understand that this is nothing less than the Obama administration's stealth implementation of the Dream Act?

The situation is even worse when you look at the way this administration has abandoned those who are trying to keep illegals out of the country.

Paul Babeu, the sheriff of Arizona's Pinal County, puts it bluntly: "What's very troubling is the fact that at a time when we in law enforcement and our state need help from the federal government, instead of sending help they put up billboard-size signs warning our citizens to stay out of the desert in my county because of dangerous drug and human smuggling and weapons and bandits and all these other things and then, behind that, they drag us into court with the ACLU."

Babeu fixed the blame directly on Obama and Holder:

> Who has partnered with the ACLU? It's the president and Eric Holder himself. And it's simply outrageous. . . . If the president would do his job and secure the border; send 3,000 armed soldiers to the Arizona border and stop the illegal immigration and the drug smuggling and the violence, we wouldn't even be in this position and where we're forced to take matters into our own hands. Our own government has become our enemy and is taking us to court at a time when we need help.[58]

Do you know how dangerous it is to work for the Border Patrol in our southwestern states?

Obama said smugly that attacks on Border Patrol agents have become "relatively stable"—at about 1,000 a year over the three years of Obama's regime.[59] That was after he had joked that Americans want our southern borders secured by "moats—with alligators."

At the same time he and Holder were conspiring to arm Mexican drug criminals, then refusing to uphold the law when criminals took over vast swaths of territory in those same border states. In fact, they're prosecuting our own Border Patrol agents instead of the drug criminals those agents are trying to arrest.

Agent Diaz was convicted of depriving one of the drug criminals he arrested in 2008 of his "rights." Despite the fact that there was no evidence against the agent, and despite the fact that he'd been exonerated of any wrongdoing by Homeland Security's Office of Inspector General and U.S.

Customs Enforcement's Office of Professional Responsibility, he was later convicted when the U.S. attorney's office for the Western District of Texas brought charges based on an internal affairs investigation. He's been sentenced to two years in prison.[60]

E-mails released by Chris Crane, president of the union that represents officers of the Immigration and Customs Enforcement (ICE) agency, to the House Judiciary Committee in early November 2011 confirmed that ICE headquarters had ordered its field agents not to arrest illegal aliens who didn't have criminal records, even if they were fugitives from justice trying to avoid being deported or who had illegally reentered the United States after having been deported. Crane testified that ICE higher-ups gave orders verbally to avoid creating a paper trail that would tie them to this activity.[61]

In the wake of its egregious refusal to uphold the laws and the Constitution of the United States, the Justice Department published a National Drug Threat Assessment in 2011.

Do you know what it said?

I'll tell you:

"Mexican-based trafficking organizations control access to the U.S.-Mexico border, the primary gateway for moving the bulk of illicit drugs into the United States. . . . The organizations control, simultaneously use, or are competing for control of various smuggling corridors that they use to regulate drug flow across the border," says the assessment. "The value they attach to controlling border access is demonstrated by the ferocity with which several rival TCOs [transnational criminal organizations] fighting over control of key corridors, or 'plazas.' "[62]

The report goes on to tell us not to expect things to change much in the next decade or so.

Do you understand what you're hearing from the traitors in the White House and the Department of Justice?

Do you realize they're thumbing their noses at the American people, especially the patriots charged with defending our borders?

Do you know that they're determined to undermine our national security by giving our southwestern states over to the drug cartels that are invading the United States?

The president's "vision" translates to lawlessness, brutality, and unspeakable violence. As one commentator explained, the "signature crimes of the most violent drug cartel in Mexico are its beheading and dismemberment of rival gang members, military personnel, law enforcement officers and public officials, and the random kidnappings and killings of civilians who get caught in its butchery and bloodletting." [63]

The administration's corrupt policies may have come full circle with another new edict issued in August 2011, when the White House announced a "community-based approach" to combating terrorism in the United States in a document called "Empowering local partners to prevent violent extremism in the United States." This announcement was followed up by a new directive from several government agencies, including the FBI, that asks managers of businesses in America to spy on their customers and report any unusual behavior to the authorities. [64]

Among the things the government wants businesses to look for are the purchase of "night flashlights" and "waterproof matches" and customers who pay with cash. They're also to be on the lookout for customers who are "missing hand/fingers" or who make "extreme religious statements" or "appear to condone violence." All of these could be indicators that the buyers are potential terrorists, according to the administration's logic.

Homeland Security Secretary Janet Napolitano, otherwise known as "Big Sis," is a Democrat loyalist with virtually no counterterrorism experience. She used the same logic when she warned, in a 2009 report titled "Right-wing Extremism: Current Economic and Political Climate Fueling Resurgence in Radicalization and Recruitment," that returning military veterans, those who favor states' rights, and people who oppose abortion should be treated as suspected terrorists. [65]

Instead of focusing on sealing our southern border, she announced that she was more concerned with white racist groups hiding in the mountains of Idaho and the sparsely populated deserts of West Texas.

The businesses informed of the new policy include gun shops, fertilizer suppliers, motels, and hotels, according to authorities.

What the White House fails to acknowledge, though, is that government institutions have routinely overlooked exactly the kind of activity it's now asking citizens to report in the interest of promoting political correctness. The U.S. Army willfully ignored the pro-jihad rants of Major Nidal Hasan and multiple reports by his superiors about the pro-Islamist Internet postings he made before Hasan, shouting "Allahu akbar," killed 13 and wounded 30 at Fort Hood, Texas, in November, 2009.

This new government edict has less to do with stopping terrorism and more to do with the Obama administration's continuing assault on our constitutional rights. It reminds me of what happens in totalitarian states everywhere as citizen is turned against citizen, family member against family member, while the state accumulates increasing power over its citizens' lives.

The first people who should be investigated under this new policy are the very people who dreamed it up: Barack Obama and the administration officials in Holder's Anti-Justice Department who are bent on extending their power over us by weakening, and ultimately eliminating, our borders and our constitutional rights.

CHAPTER 11

Crushing Obama's Cadre Before They Crush Us

It's a simple theory—one that speaks to our rugged individualism and healthy skepticism of too much government. [Trusting the markets to help us recover from a recession] fits well on a bumper sticker. Here's the problem: It doesn't work. It's never worked. It didn't work when it was tried in the decade before the Great Depression. It's not what led to the incredible post-war boom of the 50s and 60s. And it didn't work when we tried it during the last decade. . . . [What is at risk today] is whether this will be a country where working people can earn enough to raise a family, build a modest savings, own a home, and secure a retirement.

—Barack Obama, Osawatamie, Kansas, December 6, 2011

Do you understand what the president was saying in the speech he gave in Osawatamie, Kansas? Do you realize the implications of the lies and fabrications he spouted in a speech designed to conjure up the ghost of Teddy Roosevelt, the progressive president who helped push Americans further down the path of dependency and helplessness and diminishing freedom?

Under the rule of Barack Obama your freedoms have been eroded largely without your being aware of it, to the point where Americans are suddenly confronted with the fact that this president not only doesn't understand how America became the great nation it has become, he's actively promoting a political philosophy that will guarantee that we will never recover our entrepreneurial spirit and national pride.

In his Osawatamie speech, Obama essentially portrayed the U.S.—the richest, most powerful nation on earth—as a struggling soon-to-be third world country, one where individual citizens, despite our long history of valuing and promoting individualism, suddenly become a problem to be solved and not the solution to the problems.

You've gotten the message from Obama and his cadre that he must step in to control your every move and in the process strip you of your rights as an American citizen.

Do you remember Nancy Pelosi explaining that Congress had to pass health care legislation so that we could find out what was in it? As the Democrat-controlled Congress rammed that legislation through, we discovered that it was so complex even they didn't bother to read it. As a result, unless the Supreme Court overturns the abomination that this new law is, or unless a new Republican president invalidates it through executive order, you will have given over control of your medical care to a committee of bureaucrats. Your medical records will become a matter of public record, how long you live a matter of public policy.

You stood by passively as the U.S. government took increasingly aggressive and invasive measures to "protect" you.

The results?

You're now subject to dangerous and invasive scans and searches of your person every time you board a commercial flight. Minimum-wage functionaries decide—randomly, in order to make sure terrorists aren't singled out for "unfair" treatment—that anyone from young children to elderly citizens in wheelchairs is a potential threat to blow up the airliner you'll be on, regardless of ethnicity or suspicious behavior. Maybe you thought that airport security was really what these insidious practices were all about. Maybe you didn't realize that you were just being softened up so that the president could conduct invasive searches on you anywhere and at any time he wanted to, regardless of probable cause.

Thanks to the clandestinely passed NDAA legislation (signed into law on New Year's Eve!), the president of the U.S. now has the right to order you to be detained indefinitely by the U.S. military because you've been identified as posing a threat to our "national security." How long do you think it will be before being a "threat to our national security" means "speaking out against the Obama administration?"

In the interest of "national security," the president is gathering the power and the technological know-how to assume control of broadcast and Internet communications, shutting down our ability to talk with one another in case he's not happy with what we're saying.

Did you ignore his test of the national emergency broadcast network?

Did you think it was just something to be used in case of a national emergency?

His stealth takeover of our lives has now been imposed on everything from the food we eat to the fuel we need to drive our cars to the value of the money we exchange to buy these things. He's allowed his leftist apparatchiks to impose their anti-free enterprise agenda on every aspect of our lives.

Eric Holder, our Attorney General, has fought every attempt to maintain security at our borders and in our election booths. He's filed lawsuits against the public servants committed to protecting us from the ongoing criminal invasion from Mexico and South America and the onslaught of illegal aliens and unregistered Americans he's determined to see vote in the 2012 elections. Obama and Holder granted de facto amnesty to 215,000 El Salvadorans in January, 2012, using an earthquake that occurred in that country a decade ago as an excuse!

Do you remember how Obama responded when our most advanced unmanned drone went down in Iran? Despite his military and foreign policy advisors' giving him advice about how he could either destroy the drone using another unmanned aircraft or send troops in to recover or destroy it, Obama decided he didn't want to risk offending Mahmoud and the mullahs, so he let the Iranians keep it. They promptly called in the Chinese Communists to reverse engineer the aircraft, in effect giving them access to advanced American technology because our own president is afraid to confront our mortal enemies.

Did you ever consider that Obama might have made the decision he did because it was precisely the opposite of America's best interests?

Did you even entertain the idea that Barack Obama was merely continuing to level the playing field so that our Islamist and communist enemies could advance their anti-American, anti-Semitic agenda?

Do you realize that what's happening in America today amounts to nothing less than a political power-grab that resembles that of the Nazi takeover of Germany in the 1930s?

Let me acquaint you with the work of Milton Mayer, who wrote a book titled *They Thought They Were Free*. Mayer's book describes how the Nazi Party gradually assumed total control over the lives of German citizens. The Nazi takeover didn't happen in a week or a month, it happened over years. It consisted of small steps, each of which appeared innocuous enough to German citizens, but which when taken together came to represent the death knell of freedom and the rise of totalitarian control in that country.

Here is what Mayer had to say about the process:

> What happened here was the gradual habituation of the people, little by little, to being governed by surprise; to receiving decisions deliberated in secret; to believing that the situation was so complicated that the government had to act on information which the people could not understand, or so dangerous that, even if the people could understand it, it could not be released because of national security. And their sense of identification with Hitler, their trust in him, made it easier to widen this gap and reassured those who would otherwise have worried about it.
>
> This separation of government from people, this widening of the gap, took place so gradually and so insensibly, each step disguised (perhaps not even intentionally) as a temporary emergency measure or associated with true patriotic allegiance or with real social purposes. And all the crises and reforms (real reforms, too) so occupied the people that they did not see the slow motion underneath, of the whole process of government growing remoter and remoter.[1]

Can you see that the path that Obama is leading our nation down is dangerously similar to that of Nazi Germany? And that he may be using the very techniques Hitler perfected?

Let me step back here and show you what's happened in the last three years.

Let me start with the most memorable image I have taken away from the Zuccotti Park Occupy Wall Street fiasco: an older black man wearing a beret and carrying this sign around his neck: ***"Ask not what you can do for your country . . . ask what your country can do for you."***

It's Barack Obama's mantra.

It sums up the mentality not just of the OWS protesters, but of the entire Obama administration.

You might remember that it was I, Michael Savage, who defined George W. Bush as a "fiscal socialist" during the last two years of his term. I explained to you how he expanded the federal government and made it larger than the previous three administrations combined.

So if Bush was a fiscal socialist, what does that make Barack Obama?

It makes him a fiscal Marxist.

Never before in American history has a single president expanded the government so rapidly and accumulated so much power to himself. Never before have we seen this level of overt cronyism, corruption, and graft. Just as Europe is beginning to move away from large government and entitlement spending and toward lowering taxes to stimulate private enterprise, Obama, who is stuck in a perennial 1960s purple haze, is taking this nation in the opposite direction! Even Cuba is moving toward some privatization and reducing the number of Castro government jobs.

In the meantime, Barack Obama, the naked tyrant, the tired '60s radical, keeps pushing in the opposite direction.

Let me set the record straight for you about the Obama presidency with a brief catalog of his accomplishments.

He's bankrupting America with trillion-dollar-plus annual deficits—despite the fact that he hasn't submitted a budget in two years.

He stood by as millions of Americans were unfairly driven out of their homes by a robot foreclosure system.

He funneled tens of billions of stimulus dollars confiscated from taxpayers to his political donors and crony bankers to fund bailouts and green energy projects that he knew were destined to fail.

He's presided over the largest real-job loss in this country since the Great Depression.

He's waged an ongoing battle against freedom of speech and ideas contrary to his own Marxist ideology by trying to "regulate" Internet communications.

He's sent billions of dollars to energy companies in China and Brazil to promote oil exploration while at the same time doing everything in his power to shut down the American oil industry and prevent us from becoming energy independent within the next decade.

He refused for more than two years to negotiate with Iraq in order to make sure we could keep American troops in that country, and because he subsequently withdrew all of our military, he created a massive power vacuum in Iraq. In the wake of Obama's ill-advised retreat, Iran has stepped in to fill the void. We've watched as sectarian violence resurfaced and Shiite Iraqi Prime Minister Nouri al-Maliki put out an arrest warrant for his own vice president, Tareq al-Hashemi, who is a member of the rival Sunni sect. Obama's timid, apologetic foreign policy removed even the semblance of democracy that U.S. presence in Iraq helped guarantee.

Obama has presided over the most corrupt administration in history, undermining our military and giving up control of our southern border as he sold weapons to Mexican drug cartels while bringing lawsuits against states that tried to enforce immigration law.

He's bypassed Congress and the Constitution, instituting what amounts to rule by imperial fiat through executive orders and czars and cabinet secretaries who do his bidding.

He's put economic "solutions" in place that consist of taxing and spending us into indentured servitude while he transitions the U.S. into a Euro-like socialist country that fosters a dangerous dependency of citizens on the government.

He's even turned our election results over to a foreign company, a move that could compromise the way we track votes at the precinct level and eliminate our ability to challenge election results. All of our elections are now controlled by a foreign firm of questionable ownership.

Are you wondering why I call this list Obama's *accomplishments*?

Let me explain.

Barack Obama's intent is not to preside over the United States in order to guide it as it strengthens its position of economic and military pre-eminence in the world. It's just the opposite. This president of the United States is determined to undermine the values, principles, and practices on which this nation rose to greatness. He's doing that because he's none of the things he would have us believe he is.

Most Americans are asleep to this reality. The American sheeple are overwhelmed by the onslaught of lies perpetuated by so-called journalists who are nothing more than left-wing propagandists, apologists for the most radical and incompetent chief executive this country has ever been forced to endure.

Here's why I'm so worried that Obama might be re-elected.

I believe that Obama thinks his administration has sold the idea that we've weathered the worst with our economy.

Nine percent unemployment has become the new normal, and Americans have accepted the idea that we have to live with it, that it's not going to change.

Trillion-dollar deficits are now the reality as Obama plunges us deeper and deeper into debt to finance his payoffs to the American people in the form of the entitlement society. Half of all Americans now receive government benefits, and they're not likely to want to give them back.

Obama is satisfied that his economic policies have succeeded.

I'm afraid the American people are, too.

Obama has made it clear that he's not going to try to work with Congress, and he's not hiding it from the American people. Here's what he told a crowd at one of his ongoing series of campaign appearances: "We can't wait for an increasingly dysfunctional Congress to do its job. Where they won't act, I will. . . . I've told my administration to keep looking every single day for actions we can take without Congress, steps that can save consumers money, make government more efficient and responsive, and help heal the economy. And we're going to be announcing these executive actions on a regular basis."[2]

Where is the outcry against this president ignoring the separation of powers delineated in the Constitution?

Obama went so far as to laugh at Congress's ineptitude in one appearance on *The Tonight Show* with the stooge Jay Leno. He's able to do this because Congress's approval ratings have fallen as low as nine percent in opinion polling. The Republican Party itself has generated negatives over 70 percent in one poll.

Obama is going to continue to make Congress his whipping boy, and the American people aren't going to realize that what he's proposing as an alternative is that he become this nation's "glorious leader," like the tyrants he's seeking to emulate. Since the Republicans haven't been able to develop anything resembling a response, a counter strategy to provide a viable alternative to Obama's message, the president is likely to succeed here, too.

Obama advisor David Axelrod, who's back in Chicago plotting the continuing presidential takeover strategy for the 2012 election, must be smiling as he picks one Republican after another to skewer as soon as he or she begins to rise in the polls.

Among the other things our own Dictator-in-Chief believes are going to help him get reelected is the Occupy Wall Street movement. He's selling the idea that they represent 99 percent of Americans and that their "success" will rub off on him, making him look like a true populist instead of the crony capitalist thug he really is.

Again, I sense that America is going to buy this, thanks to the saturation positive news coverage the street rabble of the OWS movement are getting from the government-media complex.

Barack Obama has taken over jobs, health care, the auto industry, student loans, and mortgages. He's wiped out *habeas corpus* in the name of national security. He's eliminated the *posse comitatus* act, effectively instituting martial law by authorizing the U.S. military to arrest and detain indefinitely any U.S. citizen he and his cadre declare to be a threat.

That is why you must crush the Obama steamroller.

What I've delivered in this book is the most important analysis you'll ever read of why we must stop the tyrant in the White House now, before it's too late. If we allow Obama another four years, he will have destroyed our country. We're living under a crypto dictatorship right now. In *Trickle Down Tyranny*, I've shown you the scope of the Obama takeover and I've shown you how we can crush it before it crushes us.

ACKNOWLEDGMENTS

I want to acknowledge Greg Lewis for helping make this the book that it is, through his insights and research. Also Ian Kleinert, my literary agent, for bringing all the moving pieces together. And my eagle-eyed editor, Peter Hubbard, and deputy publisher Lynn Grady, both of William Morrow/HarperCollins, for the faith to publish this, my most seminal work.

NOTES

Chapter 1. Advice to the Next President: A Savage Worldview

1. Mackenzie Eaglen, "A Shrinking Navy," The Foundry blog, January 29, 2010, http://blog.heritage.org/2010/01/29/a-shrinking-navy/.
2. Luke Harding and Ian Traynor, "Obama abandons missile defence shield in Europe," *Guardian*, September 17, 2009, http://www.guardian.co.uk/world/2009/sep/17/missile-defence-shield-barack-obama.
3. Robert Saiget, "China's Hu urges navy to prepare for combat," Agence France-Presse, December 6, 2011, http://news.yahoo.com/chinas-hu-urges-navy-prepare-combat-160509787.html.
4. Margaret Menge, "U.S. Spent $200M on Egypt Election," *Newsmax.com*, December 2, 2011, http://www.newsmax.com/Newsfront/us-money-egypt-elections/2011/12/02/id/419829.
5. Tiffany Gabbay, "Muslim Brotherhood Takes Egypt Elections by Storm, Poised to Win Islamic Majority," *Blaze*, December 1, 2011, http://www.theblaze.com/stories/muslim-brotherhood-takes-egypt-elections-by-storm-poised-to-islamic-majority/.
6. Ibid.
7. "Britain cuts relations with Iranian banks over nuclear fear," *Business Etc.*, November 21, 2011, http://businessetc.thejournal.ie/britain-cuts-relations-with-iranian-banks-over-nuclear-fear-284918-Nov2011/.
8. "Obama urging US lawmakers to soften Iran sanctions," *Jerusalem Post*, December 6, 2011, http://www.jpost.com/International/Article.aspx?id=248396.
9. "US urges Israel to end 'isolation in the Middle East,'" BBC News, December 2, 2011, http://www.bbc.co.uk/news/world-middle-east-16014194.
10. Rob Bluey, "Defense Spending Dwindles Under Obama," The Foundry, November 20, 2011, http://blog.heritage.org/2011/11/20/chart-of-the-week-defense-spending-dwindles-under-obama/.

11. Rick Maze, "Senate committee slashes 2012 defense budget," *Air Force Times*, September 8, 2011, http://www.airforcetimes.com/news/2011/09/military-senate-committee-slashes-2012-defense-budget-090811/.

12. James Delingpole, "Climategate 2.0," *Wall Street Journal*, November 28, 2011, http://online.wsj.com/article/SB10001424052970204452104577059830626002226.html?mod=WSJ_Opinion_LEADTop.

13. "The United States of EPA," *Wall Street Journal*, November 28, 2011, http://online.wsj.com/article/SB10001424052970204630904577056393981840650.html?mod=WSJ_Opinion_AboveLEFTTop.

14. "The Non-Green Jobs Boom," *Wall Street Journal*, November 28, 2011, http://online.wsj.com/article/SB10001424052970204190704577024510087261078.html.

15. Pat Shannan, "Scientists Confirm U.S. Has World's Biggest Oil Reserves," *American FreePress.net*, June 29, 2009, http://www.americanfreepress.net/html/biggest_oil_reserves_182.html.

16. "Sovereign Bond Auction Fizzles in Germany," *Spiegel Online International*, November 23, 2011, http://www.spiegel.de/international/germany/0,1518,799550,00.html.

17. Edwin Mora, "Annual DHA Appropriations for Border Security Costs Halved Under Obama," *CNSNews.com*, November 28, 2011, http://cnsnews.com/news/article/annual-dhs-appropriations-border-security-costs-halved-under-obama.

Chapter 2. Tyranny of a Naked Marxist Presidency

1. Jonathan Turley, "The NDAA's historic assault on American Liberty," *The Guardian*, January 2, 2012, http://www.guardian.co.uk/commentisfree/cifamerica/2012/jan/02/ndaa-historic-assault-american-liberty.

2. Charlie Savage, "Secret U.S. Memo Made Legal Case to Kill a Citizen," *New York Times*, October 8, 2011, http://www.nytimes.com/2011/10/09/world/middleeast/secret-us-memo-made-legal-case-to-kill-a-citizen.html?_r=4&hp#h[].

3. Jonathan Alter, "The Obama Miracle, a White House Free of Scandal," *Bloomberg.com*, October 27, 2011, http://www.bloomberg.com/news/2011-10-27/obama-miracle-is-white-house-free-of-scandal-commentary-by-jonathan-alter.html.

4. Keith Koffler, "Michelle's Separate Travel Costs Taxpayers Thousands," White House Dossier, August 19, 2011, http://www.whitehousedossier.com/2011/08/19/michelles-separate-travel-costs-taxpayers-thousands/.

5. "Expensive massages, top shelf vodka and five-star hotels: First Lady accused of spending $10m in public money on her vacations," *Mail Online*, August 25, 2011, http://www.dailymail.co.uk/news/article-2029615/Michelle-Obama-accused-spending-10m-public-money-vacations.html.

6. Mark Tapscott, "Nearly $500K to send Michelle, family to Africa," *Washington Examiner*, October 4, 2011, http://campaign2012.washingtonexaminer.com/blogs/beltway-confidential/nearly-half-million-send-michelle-africa.

7. Bob Unruh, "U.S. sued over Michelle's secretive 'family outing,'" *WorldNetDaily*, August 25, 2011, http://www.wnd.com/?pageId=337833.

8. "Who's Footing the Bill for the Obama Family's Estimated $4 million Hawaii Vacation?" The Blaze, December 19, 2011, http://www.theblaze.com/stories/so-whos -footing-the-bill-for-the-obama-familys-4-million-hawaii-vacation/.

9. "September 11 Memorial Ceremony at the World Trade Center," C-SPAN Video Library, September 11, 2011, http://www.c-spanvideo.org/program/Ceremonyatt/start/ 2003/stop/2076.

10. "Michelle Obama's Marxist World View," *Potter Williams Report*, October 21, 2011, http://potterwilliamsreport.com/2011/10/20/michelle-obamas-marxist-world-view-3 .aspx.

11. Kirit Radia, "War Powers Showdown Heats Up in the Senate Over Libya Operations," ABC News, June 4, 2011, http://abcnews.go.com/Politics/war-powers-showdown-heats -us-libya-operations/story?id=13948848.

12. "CBS Reporter Says WH Screamed at Her Over Fast and Furious," Jammie Wearing Fool, October 4, 2011, http://jammiewearingfool.blogspot.com/2011/10/cbs-reporter -says-wh-screamed-at-her.html.

13. Joe Eaton, "Limousine Liberals? Number of government-owned limos has soared under Obama," *iWatch News*, May 31, 2011, http://www.iwatchnews.org/2011/05/31/4765/ limousine-liberals-number-government-owned-limos-has-soared-under-obama.

14. Ibid.

15. "Breaking: ALIPAC Calling for Impeachment of Barack Hussein Obama," T-Room.us, June 28, 2011, http://www.t-room.us/2011/06/breaking-alipac-calling-for-impeachment -of-barack-hussein-obama/.

16. Joe Farah, "World Net Daily: Yes, It's Time to Impeach Obama," ImpeachObama Campaign.com, March 31, 2011, http://www.impeachobamacampaign.com/worldnet -daily-yes its-time-to-impeach-obama/.

17. "Obama: Impeach and Convict Now!" Sodahead.com, March 8, 2011, http://www .sodahead.com/united-states/obama-impeach-and-convict-now/question-1567225/.

Chapter 3. Tyranny of the Government-Media Complex

1. "Tammy Haddad," *Politico*, http://www.politico.com/click/focus/tammy_haddad.html.

2. Daniel Kurtzman, "Obama at the White House Correspondents' Dinner," *Political Humor.com*, http://politicalhumor.about.com/od/barackobama/a/obama-white-house -correspondents-transcript.htm.

3. "Media Bias Basics," Media Research Center, http://www.mrc.org/biasbasics/bias basics3.asp.

4. Deborah White, "Profile of Chris Matthews of MSNBC's Hardball," US Liberal Politics, http://usliberals.about.com/od/thepressandjournalist1/p/CMatthews.htm.

5. "Amb. Marc Ginsberg," *Huffington Post*, http://www.huffingtonpost.com/amb-marc -ginsberg.

6. "Tony Blankley," *Townhall.com*, http://townhall.com/columnists/tonyblankley/.

7. "Mary Matalin," Infoplease, http://www.infoplease.com/ipa/A0880181.html.

8. "James Carville," *CNN.com*, http://www.cnn.com/CNN/anchors_reporters/carville .james.html.

9. David Theroux, "Federal Government Workers Make Twice That of Private-Sector Counterparts," *MyGovCost.com*, June 5, 2011, http://www.mygovcost.org/2011/06/05/federal-government-workers-make-twice-that-of-private-sector-counterparts/.

10. Glynnis MacNichol, "George Soros Donates $1 Million To Media Matters To Combat Fox News," *Mediaite.com*, October 20, 2010, http://www.mediaite.com/online/george-soros-donates-1million-to-media-matters-to-combate-fox-news/.

11. "George Soros to Fareed Zakaria: Fox News Uses Orwellian Newspeak and Nazi Propaganda Techniques," News Hounds, February 20, 2011, http://www.newshounds.us/2011/02/20/george_soros_to_fareed_zakaria_fox_news_uses_orwellian_newspeak_and_nazi_propaganda_techniques_.php.

12. Paul Bedard, "Report: Liberal Soros's Cash Funds 'Nonpartisan' Media Watchdogs," *U.S. News & World Report*, May 17, 2011, http://www.usnews.com/news/washington-whispers/articles/2011/05/17/report-liberal-soross-cash-funds-nonpartisan-media-watchdogs.

13. Meredith Dake, "Media Matters Declared War, Let's Give Them One," Big Government, July 11, 2011, http://biggovernment.com/mdake/2011/07/11/media-matters-declared-war-lets-give-them-one/.

14. Arthur Beesley, "Vlaams Belang blends Flemish cause into its anti-migrant agenda," *Irish Times*, April 28, 2011, http://www.irishtimes.com/newspaper/world/2011/0428/1224295619981.html.

15. "Fox Business launches series on Media Matters," *Politico*, August 4, 2011, http://www.politico.com/blogs/onmedia/0811/Fox_Business_launches_series_on_Media_Matters.html.

16. Graeme Wearden, "What has happened to Conrad Black's media empire?" *Guardian*, July 25, 2010, http://www.guardian.co.uk/media/2010/jul/26/conrad-black-media-empire-assets.

17. Brian Cathcart, "James Murdoch's Facts Trap," *Daily Beast*, July 21, 2011, http://www.thedailybeast.com/articles/2011/07/21/james-murdoch-caught-in-new-facts-from-parliament-hacking-testimony.html.

18. Sarah Rappaport, "News Of The World's Greatest Hits," *Business Insider*, July 7, 2011, http://www.businessinsider.com/news-of-the-world-stories-2011-7?op=1.

19. "What price privacy now?" Information Commissioner's Office, United Kingdom, December 13, 2006, http://www.ico.gov.uk/upload/documents/library/corporate/research_and_reports/ico-wppnow-0602.pdf.

20. Indu Chandrasekhar, Murray Wardrop, and Andy Trotman, "Phone Hacking: Timeline of Scandal," *Telegraph*, November 10, 2011, http://www.telegraph.co.uk/news/uknews/phone-hacking/8634176/Phone-hacking-timeline-of-a-scandal.html.

21. "Rupert Murdoch gets Burma-Shave pie in the face," *Los Angeles Times*, July 19, 2011, http://latimesblogs.latimes.com/showtracker/2011/07/nothing-says-answer-the-dang-question-like-a-pie-in-the-face-testimony-by-james-and-rupert-murdock-before-parliments.html.

22. "News of the World phone-hacking scandal—Friday 8 July 2011," *Guardian*, July 8, 2011, http://www.guardian.co.uk/media/blog/2011/jul/08/news-of-the-world-phone-hacking-scandal.

23. Jeff Labrecque, "Piers Morgan pulled into phone-hacking investigation by Heather Mills claim," *EW.com*, August 4, 2011, http://news-briefs.ew.com/2011/08/04/piers -morgan-heather-mills/.

24. Michael Tomasky, "U.S. Could Prosecute Murdoch," *Daily Beast*, July 19, 2011, http://www.thedailybeast.com/articles/2011/07/19/rupert-murdoch-s-american-law breaking-how-he-could-be-prosecuted.html.

25. Mark Duell and Jennifer Madison, "U.S. Justice Department launches investigation into the extent of News Corp phone hacking and bribes," *Mail Online*, July 19, 2011, http://www.dailymail.co.uk/news/article-2016529/News-World-phone-hacking-bribes -US-Justice-Department-launches-investigation.html.

26. "Why Now of All Times!" Above Top Secret, September 26, 2011, http://www.above topsecret.com/forum/thread757088/pg1.

27. Jerry Gray, "Florida Couple Are Charged In Taping of Gingrich Call," *New York Times*, April 24, 1997, http://www.nytimes.com/1997/04/24/us/florida-couple-are -charged-in-taping-of-gingrich-call.html.

28. James Risen and Eric Lichtblau, "Bush Lets U.S. Spy on Callers Without Courts," *New York Times*, December 16, 2005, http://www.nytimes.com/2005/12/16/politics/ 16program.html?pagewanted=all.

29. Eric Lichtblau and James Risen, "Bank Data Is Sifted in Secret to Block Terror," *New York Times*, June 23, 2006, http://www.nytimes.com/2006/06/23/washington/ 23intel.html?pagewanted=all.

30. David Johnston and Mark Mazzetti, "A Window Into C.I.A.'s Embrace of Secret Jails," *New York Times*, August 12, 2009, http://www.nytimes.com/2009/08/13/world/ 13foggo.html?pagewanted=all.

Chapter 4. Tyranny of the Treasurer

1. Kerry Kubilius, "Cannibalism in the Soviet Union," *East European History*, October 27, 2007, http://kerrykubilius.suite101.com/cannibalism_in_the_soviet_union-a34252.

2. "Obama Warns Against Being Self-Reliant," Walton & Johnson, October 27, 2011, http://www.waltonandjohnson.com/showarchives.html?n_id=2115.

3. Jeffrey H. Anderson, "Obama's Recovery A Flop Of Historical Proportions," *Investor's Business Daily*, September 7, 2011, http://www.investors.com/NewsAndAnalysis/ Article/583866/201109061718/Obamas-Recovery-A-Flop-Of-Historical-Proportions .htm.

4. Mortimer B. Zuckerman, "Why the Jobs Situation Is Worse Than It Looks," *U.S. News & World Report*, June 20, 2011, http://www.usnews.com/opinion/mzuckerman/ articles/2011/06/20/why-the-jobs-situation-is-worse-than-it-looks.

5. Matt Cover, "Obama Agriculture Secretary: Food Stamps Create Jobs," *CNSNews.com*, August 16, 2011, http://www.cnsnews.com/news/article/obama-agriculture-secretary -food-stamps-create-jobs.

6. "Expanding Net," http://si.wsj.net/public/resources/images/NA-BN588_Benefi_G _20111005140904.jpg.

7. Bruce Watson, "The Middle-Class Squeeze: Falling Wealth, Rising Costs," *Daily Fi-*

nance, October 27, 2011, http://www.dailyfinance.com/2011/10/27/the-middle-class
-squeeze-falling-wealth-rising-costs/?ncid=webmail30.

8. Matt Cover, "Flashback: Obama in 2008: Adding $4 Trillion to National Debt 'Un-patriotic,' " *CNSNews.com*, August 24, 2011, http://www.cnsnews.com/news/article/
flashback-obama-2008-adding-4-trillion-n.

9. Peter Roff, "A Plan Better than Obama's to Create Jobs," *U.S. News & World Report*, Sep-tember 9, 2011, http://www.usnews.com/opinion/blogs/peter-roff/2011/09/09/better
-ways-than-obamas-plan-to-create-jobs?s_cid=rss:peter-roff:better-ways-than-obamas
-plan-to-create-jobs.

10. Becket Adams, "New Jobless Claims Continue to Hover Above 400,000," *Blaze*,
October 20, 2011, http://www.theblaze.com/stories/new-jobless-claims-continue-to
-hover-above-400000/.

11. Andrew Malcolm, "961 days in, Obama becomes sick and tired of someone dawdling
about jobs," *Los Angeles Times*, August 9, 2011, http://latimesblogs.latimes.com/wash
ington/2011/09/obama-jobs-speech-congress-bad-news-polls-.html.

12. Charles Gasparino, "Wall St. Hypocrites," *New York Post*, July 29, 2011, http://www
.nypost.com/p/news/opinion/opedcolumnists/wall_st_hypocrites_rfge3Ca2Goa0FY
QU2JECcN.

13. Naftali Bendavid and Carol E. Lee, "Leaders Agree on Debt Deal," *Wall Street Jour-nal*, August 1, 2011, http://online.wsj.com/article/SB100014240531119035202045 76
480123949521268.html.

14. Stephen Dinan, "U.S. eats up most of debt limit in one day," *Washington Times*, Au-gust 3, 2011, http://www.washingtontimes.com/news/2011/aug/3/us-eats-most-debt
-limit-one-day/.

15. "U.S. Debt . . . $15 Trillion and Counting . . . ," *International Business Times*, Novem-ber 20, 2011, http://m.ibtimes.com/debt-15-trillion-counting-au-ibtimes-com-252911
.html.

16. "After Obama's Hitler Coup: The Guts To Impeach," *Executive Intelligence Review*,
August 12, 2011, http://www.larouchepub.com/other/2011/3831coup_guts_impeach
.html.

17. Bridget Johnson, "McKeon at AEI: Super Committee Deadlock Would Put U.S.
Military 'Out of Business,' " The Enterprise blog, September 12, 2011, http://blog
.american.com/2011/09/mckeon-at-aei-super-committee-deadlock-would-put-us-mili
tary-out-of-business/.

18. "Comprehensive List of Obama Tax Hikes," Americans for Tax Reform, http://www
.atr.org/comprehensive-list-obama-tax-hikes-a6433.

19. Matthew Boyle, "Nearly 20 percent of new Obamacare waivers are gourmet restau-rants, nightclubs, fancy hotels in Nancy Pelosi's district," *Daily Caller*, May 17, 2011,
http://dailycaller.com/2011/05/17/nearly-20-percent-of-new-obamacare-waivers-are
-gourmet-restaurants-nightclubs-fancy-hotels-in-nancy-pelosi%E2%80%99s-district/.

20. "Pay to Not Play: More ObamaCare Waivers Pile Up," *Redstate.com*, May 19, 2011, http://
www.redstate.com/laborunionreport/2011/05/19/pay-to-not-play-more-obamacare
-waivers-pile-up/.

21. Jeffrey Immelt, "A blueprint for keeping America competitive," *Washington Post*, Janu-

ary 21, 2011, http://www.washingtonpost.com/wp-dyn/content/article/2011/01/20/
AR2011012007089.html.

22. Matthew Saltmarsh, "U.S. Slips to Fifth Place On Competitiveness List," *New York Times*, September 7, 2011, http://www.nytimes.com/2011/09/08/business/global/us-slides-singapore-rises-in-competitiveness-survey.html.

23. Kevin Collins, "Is Immelt's Appointment Obama's Pay-Off to Red China?" *Floyd Reports*, January 27, 2011, http://floydreports.com/is-immelt-appointment-obamas-pay-off-to-red-china/.

24. "Leader of Obama Jobs Council, Jeff Immelt, Moves GE's Health Care Unit to China," *FoxNews.com*, July 25, 2011, http://nation.foxnews.com/general-electric/2011/07/25/leader-obama-jobs-council-jeff-immelt-moves-ge-s-health-care-unit-china.

25. "Obama's 'Job Czar' Jeffrey Immelt is moving X-Ray Company out of U.S. and into China," Fact Over Fiction, July 28, 2011, http://www.factoverfiction.com/article/1962.

26. "Bill Gates to Build Next-Gen Nuclear Reactors With China," FoxNews.com, December 7, 2011, http://www.foxnews.com/scitech/2011/12/07/bill-gates-to-build-next-gen-nuclear-reactors-with-china/.

27. Jake Tapper, "General Electric Paid No Federal Taxes in 2010," ABC News, March 25, 2011, http://abcnews.go.com/Politics/general-electric-paid-federal-taxes-2010/story?id=13224558.

28. Jonathan Turley, "General Electric Profits Up 77% But Company Avoids Paying a Penny in Taxes," *JonathanTurley.org*, April 21, 2011, http://jonathanturley.org/2011/04/21/general-electric-profits-up-77-but-company-avoids-paying-a-penny-in-taxes/.

29. Curtis Dubay, "U.S. to have Highest Corporate Tax Rate in the World," The Foundry blog, December 15, 2010, http://blog.heritage.org/2010/12/15/u-s-to-have-highest-corporate-tax-rate-in-the world/.

30. John McCormack, "GE Filed 57,000-Page Tax Return, Paid No Taxes on $14 Billion in Profits," *Weekly Standard*, November 17, 2011, http://www.weeklystandard.com/blogs/ge-filed-57000-page-tax-return-paid-no-taxes-14-billion-profits_609137.html.

31. Andrew S. Ross, "Bay area firms among 30 not paying federal taxes," San Francisco Chronicle, November 4, 2011, http://www.sfgate.com/cgi-bin/article.cgi?f=/c/a/2011/11/04/BUCN1LQ5IK.DTL.

32. Justin Berger, "Jobs Panel Member Whose Solar Firm Won Loan Guarantees Raises 'Conflict of Interest' Concerns," *FoxNews.com*, November 1, 2011, http://www.foxnews.com/politics/2011/11/01/jobs-panel-member-whose-solar-firm-won-2b-loan-raises-conflict-interest/.

33. "John Holdren," *FoxNews.com*, http://www.foxnews.com/topics/politics/obama-administration/john-holdren-science.htm#r_src=ramp.

34. Bob Unruh, "Congressman: Science 'czar' giving China U.S. technology," *WorldNetDaily*, November 5, 2011, http://www.wnd.com/?pageId=364197.

35. Michael J. Boskin, "The Obama Presidency by the Numbers," *Wall Street Journal*, September 8, 2011, http://online.wsj.com/article/SB10001424053111904583204576544712358583844.html?mod=WSJ_Opinion_LEADTop.

36. "Are Millionaires Taxed Less Than Secretaries?" Associated Press, September 20, 2011, http://www.cnbc.com/id/44591639/Are_Millionaires_Taxed_Less_Than_Secretaries.

37. "The Buffett Alternative Tax," *Wall Street Journal*, September 20, 2011, http://online.wsj.com/article/SB10001424053111904194604576580800735800830.html?mod=WSJ_Opinion_LEADTop.

38. "The Fed's Secret Liquidity Lifelines," *Bloomberg.com*, http://www.bloomberg.com/data-visualization/federal-reserve-emergency-lending/#/overview/?sort=nomPeakValue&group=none&view=peak&position=0&comparelist=&search=.

39. Bradley Keroun and Phil Kuntz, "Wall Street Aristocracy Got $1.2 Trillion in Secret Loans," *Bloomberg.com*, August 22, 2011, http://www.bloomberg.com/news/2011-08-21/wall-street-aristocracy-got-1-2-trillion-in-fed-s-secret-loans.html.

40. Michael Rothfeld and Jenny Strasburg, "SEC Accused of Destroying Files," *Wall Street Journal*, August 19, 2011, http://online.wsj.com/article/SB1000142405311190407060457651484017105421.6.html.

41. Peter Ferrara, *America's Ticking Bankruptcy Bomb* (New York: Broadside Books, 2011), pp. 13–14.

42. Addison Wiggin, "The next financial crisis will be hellish, and it's on its way," *Forbes*, November 16, 2011.

43. Gerard Baker, "Why Barack Obama picked a political Who's Who," *Sunday Times*, December 5, 2008, http://www.timesonline.co.uk/tol/comment/columnists/article5288780.ece.

44. "Obama Adviser Robert Rubin Sued For Citi Ponzi Scheme," Investment Watch, January 11, 2009, http://investmentwatchblog.com/obama-adviser-robert-rubin-sued-for-citi-ponzi-scheme/.

45. Steven Malanga, "Other people's money," *New York Post*, November 23, 2011, http://www.nypost.com/p/news/opinion/opedcolumnists/other_people_money_bnv7Yjd8IAli3dAJfSbK7M.

46. "The Good, the Bad, and the Ugly—Part Three," Burning Platform, May 16, 2011, http://www.theburningplatform.com/?tag=great-depression.

47. "H.R. 4173—Dodd-Frank Wall Street Reform and Consumer Protection Act," Open Congress, http://www.opencongress.org/bill/111-h4173/show.

48. Andrew Leonard, "The Dodd-Frank bank reform bill: A deeply flawed success," *Salon.com*, June 25, 2010, http://www.salon.com/technology/how_the_world_works/2010/06/25/the_dodd_frank_bank_reform_bill.

49. Edwin Mora, "TARP Watchdog Casts Doubt on Financial 'Reform' Law's Promise to End Bailouts," *CNSNews.com*, January 28, 2011, http://www.cnsnews.com/news/article/tarp-watchdog-cast-doubts-dodd-frank-bil.

50. "Fiat Money History in the US," http://www.kwaves.com/fiat.htm.

51. "Alan Greenspan: We Can Always Print More Money," YouTube, August 7, 2011, http://www.youtube.com/watch?v=q6vi528gseA.

52. "Zimbabwe removes 12 zeros from its currency," CNN, http://edition.cnn.com/2009/WORLD/africa/02/02/zimbabwe.dollars/index.html.

53. Gregor MacDonald, "Paper vs. Real: Exit From Normal, Ecological Economics, and Probabilistic Regimes in One Chart," http://gregor.us; "Suicide Banking," *Keiser*

Report, May 5, 2011, http://maxkeiser.com/2011/05/10/keiser-report-economic
-euthanasia/.

54. Conor Dougherty, "Income Slides to 1996 Levels," *Wall Street Journal*, September 14,
2011, http://online.wsj.com/article/SB10001424053111904265504576568543968213
896.html?KEYWORDS=income+slides+to+1996+levels.

55. "US Poverty Hits Record High: 1 in 6 Americans Living Below Poverty Line," Economy Watch, September 15, 2011, http://www.economywatch.com/in-the-news/us
-poverty-hits-record-high-1-in-6-americans-living-below-poverty-line.15-09.htm.

56. Vivien Lou and Joshua Zumbrun, "Fed Presidents Signal Employment Growth
Too Slow to Remove Record Stimulus," Bloomberg News, May 5, 2011, http://bx
.businessweek.com/federal-reserve/view?url=http%3A%2F%2Fnoir.bloomberg.com%
2Fapps%2Fnews%3Fpid%3D20601068%26sid%3Dadf3gjeXmCVE.

57. Ian Cowie, "Why our purchasing power is set to suffer the biggest squeeze since 1870,"
Telegraph, May 3, 2011, http://blogs.telegraph.co.uk/finance/ianmcowie/100010094/
why-our-purchasing-power-is-set-to-suffer-the-biggest-squeeze-since-1870/.

58. Sara Murray, "About 1 in 7 in U.S. Receive Food Stamps," *Wall Street Journal*, May 3,
2011, http://blogs.wsj.com/economics/2011/05/03/about-1-in-7-americans-receive-food
-stamps/.

59. Matt Krantz, "Silver price drops, sending metal into a bear market," *USA Today*,
May 5, 2011, http://www.usatoday.com/money/markets/2011-05-05-silver-bear-mar
ket_n.htm.

60. Jonathan Stempel, "JPMorgan, HSBC sued for alleged silver conspiracy," Reuters,
October 27, 2010, http://www.reuters.com/article/2010/10/27/us-jpmorgan-hsbc
-lawsuit-idUSTRE69Q5HQ20101027.

61. "Economic Euthanasia," *Keiser Report*, May 10, 2011, http://maxkeiser.com/2011/
05/10/keiser-report-economic-euthanasia/.

62. "Hotspots with Max Keiser," YouTube, April 19, 2011, http://www.youtube.com/
watch?v=A5zXU1bQ3tQ.

63. "Europe Probes Goldman Sachs, JPMorgan Investment Banks Over Default Swaps,"
All Voices, April 29, 2011, http://www.allvoices.com/contributed-news/8933797
-europe-probes-goldman-sachs-jpmorgan-investment-banks-over-default-swaps.

64. Louise Story, Landon Thomas, and Nelson D. Schwartz, "Wall St. Helped to Mask
Devt Fueling Europe's Crisis," *New York Times*, February 13, 2010, http://www.ny
times.com/2010/02/14/business/global/14debt.html.

65. Elena Moya, "Europe bars Wall Street banks from government bond sales," *Guardian*, March 8, 2010, http://www.guardian.co.uk/business/2010/mar/08/us-banks
-european-bond-trading.

66. Elena Moya, "Europe freezes out Goldman Sachs," *Guardian*, July 18, 2010, http://
www.guardian.co.uk/business/2010/jul/18/goldman-sachs-europe-sovereign-bond-sales.

67. Brian Blackstone, "Banker's Exit Rattles Markets," *Wall Street Journal*, September 10,
2011, http://online.wsj.com/article/SB10001424053111903285704576560411990091
924.html?KEYWORDS=banker%27s+exit+rattles+markets.

68. "Biggest Holders of US Government Debt," CNBC News, http://www.cnbc.com/
id/29880401/The_Biggest_Holders_of_US_Government_Debt?slide=16.

69. Matt Jarzemsky, "S&P Cuts U.S. Ratings Outlook to Negative," *Wall Street Journal*, April 18, 2011.

70. Damian Paletta and E. S. Browning, "U.S. Warned on Debt Load," *Wall Street Journal*, April 19, 2011, http://online.wsj.com/article/SB100014240527487040040045762 70693061767996.html.

71. "United States of America Long-Term Rating Lowered to 'AA+' Due to Political Risks, Rising Debt Burden; Outlook Negative," Standard & Poor's, August 5, 2011, http://www.standardandpoors.com/ratings/articles/en/us/?assetID=1245316529563.

72. Jim Hoft, "Obama Created or Saved 450,000 Government Jobs, Lost 1,000,000 Private Sector Jobs With Stimulus," Gateway Pundit, May 16, 2011, http://gatewaypundit.rightnetwork.com/2011/05/obama-created-or-saved-450000-government-jobs-lost-1000000-private-sector-jobs-with-stimulus/.

73. Paul Haven, "New law will let Cubans buy and sell real estate," Associated Press, November 3, 2011, http://www.miamiherald.com/2011/11/03/2485009/cuba-legalizes-sale-purchase-of.html#ixzz1ckfpwhy5.

74. Zhou Xin, Simon Rabinovitch, and Kevin Yao, "U.S. in Worse Shape than Europe: China," Midas Letter, December 9, 2010, http://www.midasletter.com/index.php/u-s-in-worse-shape-than-europe-china/.

75. "China Joins Russia in Blasting U.S. Borrowing," Bloomberg News, August 3, 2011, http://www.bloomberg.com/news/2011-08-03/china-s-zhou-to-monitor-u-s-debt-as-xinhua-sees-bomb-yet-to-be-defused.html.

76. "Major Foreign Holders of Treasury Securities," U.S. Treasury, http://www.treasury.gov/resource-center/data-chart-center/tic/Documents/mfh.txt.

77. Martin Crutsinger, "China trims holdings of US Treasury debt," AP, December 15, 2011, http://economictimes.indiatimes.com/news/international-business/china-trims-holdings-of-us-treasury-debt/articleshow/11123611.cms.

78. Phil Gramm, "The Obama Growth Discount," *Wall Street Journal*, April 15, 2011, http://online.wsj.com/article/SB10001424052748703983104576262763594126624.html?mod=ITP_opinion_0.

79. Ibid.

80. "The Presidential Divider," *Wall Street Journal*, April 14, 2011, http://online.wsj.com/article/SB10001424052748703730104576260911986870054.html.

Chapter 5. Tyranny of Obama's Radical Accomplices

1. "The Buffett 391,000," *Wall Street Journal*, October 14, 2011, http://online.wsj.com/article/SB10001424052970203914304576629302662250680.html?mod=WSJ_Opinion_AboveLEFTTop.

2. "Hedge funds raked in $6.1 billion in August," Reuters, October 10, 2011, http://www.rawstory.com/rs/2011/10/10/hedge-funds-raked-in-6-1-billion-in-august/.

3. "Are You Smarter Than a Wall Street Occupier?" *New York*, October 19, 2011, http://nymag.com/daily/intel/2011/10/occupy_wall_street_quiz.html.

4. Charles Gasparino, "Stockholm Syndrome," *New York Post*, October 20, 2011, http://www.nypost.com/p/news/opinion/opedcolumnists/stockholm_syndrome_zQj3ht17 c8dFeAXtVKXhKI.

5. "#OccupyWallStreet Insiders: 'Main political goal overthrow the gov't,' " Ironic Surrealism, October 11, 2011, http://ironicsurrealism.com/2011/10/11/occupywallstreet -leaders-main-political-goal-overthrow-the-govt-lot-of-the-protesters-anarchists-revolu tionaries/.

6. Lachlan Cartwrightand Bob Fredericks, "Sex, drugs and hiding from the law at Wall Street protests," *New York Post*, October 10, 2011, http://www.nypost.com/p/news/ local/manhattan/it_nyc_lam_sterdam_bmE4vlV5aDUWhBRv9IbaiK#ixzz1aNS6Loiy.

7. Stella Paul, "When America's Children Come Home to Roost," *American Thinker*, October 14, 2011, http://www.americanthinker.com/2011/10/americas_children_come _home_to_roost.html.

8. Aaron Klein, "Look who's behind Occupy Jew-hating," *WorldNetDaily*, November 7, 2011, http://www.wnd.com/?pageId=365377.

9. Ryan Jones, "Israeli worried by anti-Semitic flavor of 'Occupy Wall Street' protests," *Israel Today*, October 16, 2011, http://www.israeltoday.co.il/News/tabid/178/nid/ 22978/language/en-US/Default.aspx.

10. "New gaffe: Obama confuses Jews with janitors," *Los Angeles Times*, September 26, 2011, http://latimesblogs.latimes.com/washington/2011/09/obama-congressional-black -caucus-video-gaffe.html.

11. Bob Unruh, "Unmasked! 'Occupy' protests sending sharp anti-Jew message," *WorldNetDaily*, October 20, 2011, http://www.wnd.com/index.php?fa=PAGE.view&pageId =358449.

12. "Proposed list of OWS demands," Occupy Wall Street, October 23, 2011, http:// occupywallst.org/forum/proposed-list-of-ows-demands/.

13. Perry Chiaramonte, "Occupy Protests Plagued by Reports of Sex Attacks, Violent Crime," *FoxNews.com*, November 2, 2011, http://www.foxnews.com/us/2011/11/09/ rash sex attacks-and-violent-crime-breaks-out-at-occupy-protests/.

14. Matthew Kuruvila and Demian Bulwa, "Man shot to death near Occupy Oakland Camp," *San Francisco Chronicle*, November 11, 2011, http://www.sfgate.com/cgi-bin/ article.cgi?f=/c/a/2011/11/10/BA05lLTHDK.DTL.

15. Kevin Fasick, Sally Goldenberg, and Bob Fredericks, "Protesters force café layoffs as biz drops," *New York Post*, November 2, 2011, http://www.nypost.com/p/news/local/ manhattan/real_job_killers_2bVY2PIRGWUrVIqREYgtoL.

16. Ikimulisa Livingston, Hannah Rappleye, and Lorena Mongelli, "OWS protesters cause mayhem across the city, 275 arrested," *New York Post*, November 18, 2011, http:// www.nypost.com/p/news/local/manhattan/roughly_people_demonstration_gathered _dqucDJsloJYCdedLgh4rwL.

17. "Obama's Army," *WorldNetDaily*, October 27, 2011, http://www.wnd.com/?pageId=3 59809#ixzz1cpR9piVA.

18. "Soros Money Is Behind Occupy Wall Street Protests," Reuters, October 13, 2011, http://www.thegatewaypundit.com/2011/10/reuters-soros-money-behind-occupy-wall -street-protests/.

19. Jana Winter, "ACORN Officials Scramble, Firing Workers and Shredding Documents, After Exposed as Players Behind Occupy Wall Street Protests," *FoxNews.com*, November 3, 2011, http://www.foxnews.com/us/2011/11/03/acorn-officials-scramble -firing-workers-and-shredding-documents-after-exposed/.

20. Meredith Somers, "With winter looming, labor unions fortify Occupy camp," *Washington Times*, November 16, 2011, http://www.washingtontimes.com/news/2011/nov/16/unions-back-occupy-dc/.

21. John McCormack, "Pelosi on Occupy Wall Street Protesters: 'God Bless Them,' " *Weekly Standard*, October 6, 2011, http://www.weeklystandard.com/blogs/pelosi-occupy-wall-street-protesters-god-bless-them_595117.html.

22. "National Air and Space Museum Closed After Protesters Storm Building," Associated Press, http://cnsnews.com/news/article/national-air-and-space-museum-closed-after-protesters-swarm-building.

23. "Union Violence: Longshoremen Storm Port, Damage Railroad," Big Government, September 8, 2011, http://biggovernment.com/publius/2011/09/08/union-violence-longshoremen-storm-port-damage-railroad/.

24. Christopher Rugaber, "Economy adds 103,000 jobs, but it's not enough," *Palm Beach Post*, October 7, 2011, http://www.palmbeachpost.com/money/economy-adds-103-000-jobs-but-its-not-1900817.html.

25. Ed Morrissey, "Obama says Wisconsin conducting 'an assault on unions,' " Hot Air, February 17, 2011, http://hotair.com/archives/2011/02/17/obama-says-wisconsin-conducting-an-assault-on-unions/.

26. "Given Half a Chance: The Schott 50 State Report on Public Education and Black Males," Schott Foundation for Public Education, 2008, http://www.blackboysreport.org/files/schott50statereport-execsummary.pdf.

27. "Obama . . . Joins Hands With His Fellow-Communist Thug Richard Trumka From AFL-CIO On Labor Day," September 16, 2010, http://itmakessenseblog.com/2010/09/16/obama-joins-hands-with-his-fellow-communist-thug-richard-trumka-from-afl-cio-on-labor-day/.

28. "Is the White House Helping to Orchestrate a National Union Uprising? You be the Judge," *Redstate.com*, February 22, 2011, http://www.redstate.com/laborunionreport/2011/02/22/is-the-white-house-helping-to-orchestrate-a-national-union-uprising-you-be-the-judge/.

29. Webutante, "The Late Saul Alinsky, Obama's Radical Community Organizer Inspirer-in-Chief," Webutante, November 17, 2008, http://webutante07.blogspot.com/2008/11/saul-alinsky-obamas-mentor-in-chief.html.

30. Alan Caruba, "Applying Alinsky: Why Obamacare Makes No Sense," *Canada Free Press*, February 26, 2010, http://www.canadafreepress.com/index.php/article/20448.

31. "Communists boast of 'mentoring' Axelrod," *WorldNetDaily*, February 16, 2011, http://www.wnd.com/?pageId=151877.

32. Patrick T. Reardon, "The Agony and the Agony," *Chicago Tribune*, June 24, 2007, http://www.chicagotribune.com/news/local/chi-070620axelrod-htmlstory,0,7217326.htmlstory.

33. Howard Fineman, "Axelrod's Alinsky-ism gave me an AH-HA! moment today," *Huffington Post*, November 11, 2010, http://rartee.wordpress.com/2010/11/11/david-axelrods-alinsky-ism-gave-me-and-ah-ha-moment-today/.

34. "Interview with Saul Alinsky, Part Two," Progress Report, http://www.progress.org/2003/alinsky3.htm.

35. Shayna Stron, "An Internet Guide to Community Organizing," Poverty & Race Research Action Council, http://www.prrac.org/full_text.php?text_id=964&item_id=8814& newsletter_id=0&header=Community+Organizing.

36. Ibid., p. 52.

37. Ibid.

38. John Hawkins, "The Best Quotes From Saul Alinsky's 'Rules For Radicals,' " Rightwing News, http://rightwingnews.com/quotes/the-best-quotes-from-saul-alinskys-rules-for -radicals/.

39. John Leo, "Islamic Radicals Take Advantage of Western Liberalism," Real Clear Politics, February 12, 2006, http://www.realclearpolitics.com/Commentary/com-2_12_06 _JL.html.

40. Rob Port, "Chris Matthews: Mark Levine, Michael Savage Are To Blame For Tuscon Shooting," *SayAnythingBlog.com*, January 11, 2011, http://sayanythingblog.com/entry/ chris-matthews-mark-levine-michael-savage-are-to-blame-for-tucson-shooting/.

41. "Saul Alinsky's son: 'Obama learned his lesson well,' " *Canada Free Press*, September 2, 2008, http://www.canadafreepress.com/index.php/article/4784.

Chapter 6. Tyranny of Obama's Corporate Cronies

1. "President Obama's 'Czars,' " *Politico*, September 8, 2009, www.politico.com/news/ stories/0909/26779.html.

2. Jeremy Arias, "$1000 Fine Cut in Takoma Park Tree Case," *Washington Post*, November 26, 2009, http://www.washingtonpost.com/wp-dyn/content/article/2009/11/25/ AR2009112502482.html.

3. "City of Takoma Park—Tree Protection or Removal," Tree World, http://www.tree world.info/f14/takoma-park-maryland-usa-tree-protection-11499.html.

4. Arias, "$1000 Fine Cut in Takoma Park Tree Case."

5. Ibid.

6. Scott London, "Book Review: Democracy and the Problem of Free Speech," http:// www.scottlondon.com/reviews/sunstein.html.

7. "Cass Sunstein Quotes," http://www.stopsunstein.com/media/pdf/Sunstein%20quote %20file.pdf.

8. London, "Book Review: Democracy and the Problem of Free Speech by Cass R. Sunstein."

9. "Obama's unpaid 'pay czar' got $120K annual salary UPDATED! He gave it back, but still has to pay taxes on it," *Washington Examiner*, September 9, 2010, https:// washingtonexaminer.com/blogs/beltway-confidential/obama039s-unpaid-039pay-czar 039-got-120k-annual-salary-updated-he-gave-it-.

10. Jeffrey H. Anderson, "Obama Insists on Funding 'Czars,' After Agreeing to Cut Funds for 'Czars,' " *Weekly Standard*, April 18, 2009, http://www.weeklystandard.com/blogs/ obama-insists-funding-czars-after-agreeing-cut-funds_557545.html.

11. Robin Bravender, "Budget deal axes 'czars' already gone," *Politico*, April 12, 2011, http://www.politico.com/news/stories/0411/53001.html.

12. "Statement by the President on H.R. 1473," White House, April 15, 2011, http:// www.whitehouse.gov/the-press-office/2011/04/15/statement-president-hr-1473.

13. "United States Constitution—Article 2, Section 2," U.S. Constitution Online, http://www.usconstitution.net/xconst_A2Sec2.html.

14. Alexander Hamilton, "The Federalist No. 69: The Real Character of the Executive," Federalist Papers, March 14, 1788, http://www.constitution.org/fed/federa69.htm.

15. Ibid.

16. Isabel Vincent and Melissa Klein, "Holder's 4G tax lax," *New York Post*, April 17, 2011, http://www.nypost.com/p/news/local/queens/holder_tax_lax_ZLrjDOZehZHon7f DakSwTL.

17. Ibid.

18. Jonathan Weisman, "Geithner's Tax History Muddles Confirmation," *Wall Street Journal*, January 14, 2009, http://online.wsj.com/article/SB123187503629378119.html.

19. Brian Wingfield, "Geithner's Tax Troubles Are Serious," *Forbes*, January 13, 2009, http://www.forbes.com/2009/01/13/treasury-geithner-obama-biz-beltway-cx_bw_0113geithner 2.html.

20. Weisman, "Geithner's Tax History Muddles Confirmation."

21. Alex Spillius, "Geithner's tax evasion," *Telegraph*, January 22, 2009, http://blogs.tele graph.co.uk/news/alexspillius/8174427/Tim_Geithners_tax_evasion_/.

22. Diana West, "Give Panetta the Pinko Slip," *TownHall.com*, June 17, 2011, http://townhall.com/columnists/dianawest/2011/06/17/give_panetta_the_pinko_slip/page/full/.

23. Ibid.

24. Aaron Klein, "Obama's DoD nominee linked to Cold War supporters of Soviets," *WorldNetDaily*, June 16, 2011, http://www.wnd.com/?pageId=311485.

25. Irene North, "Why the TSA won't improve its image anytime soon," Daily Censored, April 25, 2011, http://dailycensored.com/2011/04/25/why-the-tsa-wont-improve-its -image-anytime-soon/.

26. Anne E. Kornblut, "White House moving to repair troubled relationship with cabinet," *Washington Post*, March 9, 2011, http://www.washingtonpost.com/wp-dyn/content/article/2011/03/08/AR2011030805751.html.

27. Ibid.

28. "Press Briefing on the BP Oil Spill on the Gulf Coast," The White House, Office of the Press Secretary, April 29, 2010, http://www.whitehouse.gov/the-press-office/press -briefing-bp-oil-spill-gulf-coast.

29. Anne E. Kornblut, "White House moving to repair troubled relationship with cabinet."

30. "Obama Fails To Meet With Six Cabinet Members For Two Years, But Talks or Meets With AFL-CIO President Almost Daily," SodaHead, February 23, 2011, http://www .sodahead.com/united-states/obama-fails-to-meet-with-six-cabinet-members-for-two -years-but-talks-or-meets-with-afl-cio-presiden/question-1535985/.

31. David Shepherdson and Christina Rogers, "GM's Akerson Pushing for Higher Gas Taxes," *Detroit News*, June 7, 2011, http://newmediajournal.us/indx.php/item/1796.

32. Ibid.

33. "Alfred Dreyfus and 'The Affair,'" Jewish Virtual Library, http://www.jewishvirtual library.org/jsource/anti-semitism/Dreyfus.html.

34. Jeff Jacoby, "When Zola wrote 'J'accuse!' " *Boston Globe*, March 30, 2008, http://www
.boston.com/bostonglobe/editorial_opinion/oped/articles/2008/03/30/when_zola
_wrote_jaccuse/.

35. Michael Savage, "From the Dreyfus affair to the Strauss-Kahn affair," *Michael
Savage.com*, http://www.michaelsavage.wnd.com/2011/07/savage-from-the-dreyfus
-affair-to-the-strauss-kahn-affair/.

36. Shelley Ross, "Strauss-Kahn Fallout: NY DA Takes Five Steps Backwards For Rape
Accusers," *Huffington Post*, July 6, 2011, http://www.huffingtonpost.com/shelley-ross/
strausskahn-fallout-ny-da_b_888803.html.

37. John Swaine, "Dominique Strauss-Kahn Walks Free After Maid Rape Case Crum-
bles," *Telegraph*, July 1, 2011, http://www.telegraph.co.uk/finance/dominique-strauss
-kahn/8611957/Dominique-Strauss-Kahn-walks-free-after-maid-rape-case-crumbles
.html.

38. Jim Dwyer and Michael Wilson, "Strauss-Kahn Accuser's Call Alarmed Prosecutors," *New York Times*, July 1, 2011, http://www.nytimes.com/2011/07/02/nyregion/
one-revelation-after-another-undercut-strauss-kahn-accusers-credibility.html?page
wanted=all.

39. Brad Hamilton and Larry Celona, "Maid 'Laid' Low as DA Paid for Digs," *New York
Post*, July 3, 2011, http://www.nypost.com/p/news/local/manhattan/she_laid_low
_as_da_paid_for_digs_8Udq6nhQaHaC4KOOfkctpI.

40. Dwyer and Wilson, "Strauss-Kahn Accuser's Call Alarmed Prosecutors."

41. Ibid.

42. Hamilton and Celona, "Maid 'Laid' Low as DA Paid for Digs."

43. John Eligon, "Strauss Kahn Is Released as Case Teeters," *New York Times*, July 1, 2011,
http://www.nytimes.com/2011/07/02/nyregion/new-yorkers-and-french-await-latest
-dominique-strauss-kahn-legal-turn.html?pagewanted=all.

44. Alan Katz, "Strauss-Kahn Divides French on Political Return, *Parisien* Says," *Bloomberg
.com*, July 3, 2011, http://www.bloomberg.com/news/2011-07-03/strauss-kahn-divides
-french-on-political-return-parisien-says.html.

45. Michael Weissenstein and Angela Charlton, "Tristane Banon to File Sexual Assault
Lawsuit Against Dominique Strauss-Kahn," *Huffington Post*, July 4, 2011, http://www
.huffingtonpost.com/2011/07/04/tristane-banon-dominique-strauss-kahn-dsk_n_88
9684.html.

46. "Christine Lagarde: Executive Profile," *Bloomberg Businessweek*, http://investing.business
week.com/businessweek/research/stocks/people/person.asp?personId=160171&ticker
=INGA:NA&previousCapId=874263&previousTitle=Dexia%20Bank%20Belgium
%20SA.

47. Julian Gavaghan, "Synchronized swimming, flirting for France and a VERY risqué
dominatrix drawing: The secret life of new IMF boss Christine Lagarde," *Mail Online*,
July 1, 2011, http://www.dailymail.co.uk/news/article-2009973/Christine-Lagarde
-Secret-life-new-IMF-boss.html.

48. Ibid.

49. Fred Schulte, John Aloysius, and Jeremy Borden, "Obama Rewards Big Bundlers with
Jobs, Commissions, Stimulus Money, Government Contracts, and More," *iWatch
News*, June 15, 2011, http://www.iwatchnews.org/node/4880.

50. Derek Kravitz, "Tapping the Governor's Phone," *Washington Post*, December 9, 2008, http://voices.washingtonpost.com/washingtonpostinvestigations/2008/12/good_stuff _for_the_people_of_i.html.

51. Schulte, Aloysius, and Borden, "Obama Rewards Big Bundlers with Jobs, Commissions, Stimulus Money, Government Contracts, and More."

52. Ibid.

53. Sam Stein, "Obama Campaign Advised White House Staff to Give Top Donor Sense of Access," *Huffington Post*, June 13, 2011, http://www.huffingtonpost.com/ 2011/06/13/obama-campaign-arm-advise_n_875986.html.

54. Ibid.

55. David Willman, "Cost, need questioned in $433-million smallpox deal," *Los Angeles Times*, November 13, 2011, http://www.latimes.com/news/nationworld/nation/la-na -smallpox-20111113,0,6456082,full.story.

Chapter 7. Tyranny of the Egghead Wars

1. Kristinn Taylor, "Obama: Our Partyboy President," Big Government, March 17, 2011, http://biggovernment.com/ktaylor/2011/03/17/obama-our-partyboy-president/.

2. Julianna Goldman and Iuri Dantas, "Obama Tells Rousseff U.S. to Be Among Brazil's 'Best Customers,' " *Bloomberg Businessweek*, March 19, 2011, http://www.businessweek .com/news/2011-03-19/obama-tells-rousseff-u-s-to-be-among-brazil-s-best-customers -.html.

3. "Obama: 'We cannot stand idly by,' " *Telegraph*, March 20, 2011, http://www.tele graph.co.uk/news/worldnews/africaandindianocean/libya/8393482/Barack-Obama-we -cannot-stand-idly-by.html.

4. Robert Dreyfuss, "Obama's Women Advisers Pushed War Against Libya," Fox News, March 20, 2011, http://nation.foxnews.com/barack-obama/2011/03/20/obamas -women-advisers-pushed-war-against-libya.

5. Lee, Tim, "Sunstein on the Second Amendment," The American Scene, November 4, 2007, http://theamericanscene.com/2007/11/14/sunstein-on-the-second-amendment.

6. Brent Lang, "Power, Who Called Clinton 'Monster,' Joins Obama White House," CBS News, January 30, 2009, http://www.cbsnews.com/8301-503544_162-4764495 -503544.html.

7. "Insider Report: Hillary's College Thesis Revealed," *Newsmax.com*, March 4, 2007, http://archive.newsmax.com/archives/articles/2007/3/4/184310.shtml.

8. "US War Criminals: Where are they now? Madeleine Albright," Kadaitcha, January 16, 2011, http://www.kadaitcha.com/2011/01/16/us-war-criminals-where-are-they -now-madeleine-albright/.

9. Steve Dennis, "The Obama Doctrine: Is Ivory Coast next?" America's Watchtower, April 3, 2011, http://americaswatchtower.com/2011/04/03/the-obama-doctrine-is -ivory-coast-next/.

10. "About Crisis Group," International Crisis Group, http://www.crisisgroup.org/en/ about.aspx.

11. "Gates: Libya not 'a vital interest' for US, but part of region that's of vital interest,"

Washington Post, March 27, 2011, http://www.washingtonpost.com/gates-libya-not-a -vital-interest-for-us-but-part-of-region-thats-of-vital-us-interest/2011/03/27/AF3ZcUiB _story.html.

12. Nicholas Ballasy, "Rangel: Obama Needed Congressional Authorization for Military Intervention in Libya," *CNSNews.com*, March 31, 2011, http://www.cnsnews.com/ news/article/rangel-obama-did-not-have-constitutional.

13. "United States Ending Its Air Support of No-Fly Zone in Libya, Leaving NATO to Take Reins," Fox News, April 1, 2011, http://www.foxnews.com/us/2011/04/01/ ending-air-combat-role-libya/.

14. Pauline Jelinek, "CIA's presence doesn't break no-troops promise," *Fort Worth Star-Telegram*, March 31, 2011, http://www.star-telegram.com/2011/03/31/2965784/cias -presence-doesnt-break-no.html.

15. Hadeel Al-Shalchi, "Fighting's effect on Libya civilians remains murky," *Los Angeles Daily News*, March 25, 2011, http://www.dailynews.com/politics/ci_17698814.

16. Thom Shanker and Charlie Savage, "NATO Warns Rebels Against Attacking Civilians," *New York Times*, April 1, 2011, http://www.nytimes.com/2011/04/01/world/ africa/01civilians.html.

17. "Text of Ghadafi letter to Obama," Yahoo News, April 6, 2011, http://news.yahoo .com/s/ap/20110406/ap_on_re_us/us_us_libya_text.

18. Karen Hawkins, "Farrakhan: Libya has lent Nation of Islam millions," Yahoo News, March 31, 2011, http://news.yahoo.com/s/ap/20110401/ap_on_re_us/us_nation_of _islam_farrakhan.

19. "US hopes military action will result in Gaddafi quitting," *One India News*, April 2, 2011, http://news.oneindia.in/2011/04/02/ushopes-military-action-will-result-in -gaddafiquitting-aid0126.html.

20. "Dr. ElBaradei Calls for 'Timeout' on Iran Nuclear Issue," International Atomic Energy Agency, January 29, 2007, http://www.iaea.org/newscenter/news/2007/iran _timeout.html.

21. Danielle Pletka and Michael Rubin, "ElBaradei's Real Agenda," *Wall Street Journal*, February 25, 2008, http://online.wsj.com/article/SB120389990086289395.html.

22. Ibid.

23. Wes Barrett, "Group Says Spies Have Found Secret Iranian Nuclear Facility," Fox News, April 7, 2011, http://www.foxnews.com/world/2011/04/07/group-says-spies -secret-iran-nuclear-facility/.

24. Tim Ross, Matthew Moore, and Steven Swinford, "Egypt's protests: America's secret backing for rebel leaders behind uprising," *Telegraph*, January 28, 2011, http://www .telegraph.co.uk/news/worldnews/africaandindianocean/egypt/8289686/Egypt-protests -Americas-secret-backing-for-rebel-leaders-behind-uprising.html.

25. Ibid.

26. Charles Levinson and Steve Stecklow, "Inside Opposition's War Room," *Wall Street Journal*, February 3, 2011, p. 1.

27. Andrew Lee Butters, "Is Egypt's Parliament Finally Ready for Prime Time?" *Time*, January 31, 2011, http://www.time.com/time/world/article/0,8599,2045444,00.html.

28. Jonathan Tirone, "Iran Divides Mubarak's Troubleshooter, Opposition's ElBaradei,"

Bloomberg.com, February 2, 2011, http://www.bloomberg.com/news/2011-02-02/iran -divides-mubarak-s-troubleshooter-opposition-s-elbaradei.html.

29. Richard A. Oppel, "Foreign Fighters in Iraq Are Tied to Allies of U.S.," *New York Times*, November 22, 2007, http://www.nytimes.com/2007/11/22/world/middleeast/ 22fighters.html.

30. Gordon Rayner, "Rebels argued over whether to kill Gaddafi as he begged for his life," *Telegraph*, October 21, 2011, http://www.telegraph.co.uk/news/worldnews/africaand indianocean/libya/8842021/Rebels-argued-over-whether-to-kill-Gaddafi-as-he-begged -for-his-life.html.

31. "Moammar Kadafi's family reportedly will sue NATO," *Los Angeles Times*, October 26, 2011, http://latimesblogs.latimes.com/world_now/2011/10/moammar-kadafi -family-nato.html.

32. Ruth Sherlock, "Libya revolutionaries burn, loot village homes," *Los Angeles Times*, October 5, 2011, http://articles.latimes.com/2011/oct/05/world/la-fg-libya-fighting -20111006.

33. Kimberly Dozier and Lolita C. Baldor, "Crash kills members of SEAL Team 6," Associated Press, August 6, 2011, http://abcnews.go.com/Politics/wireStory?id=1424 6233.

34. Allan Hall, "Dozens of U.S. paratroopers injured after parachute jump during mock battle goes horrifically wrong," *Daily Mail*, October 7, 2011, http://www.dailymail .co.uk/news/article-2046471/Dozens-U-S-paratroopers-injured-mock-battle-Slovakians -goes-horrifically-wrong.html.

35. "Obama urges debt panel to reach for deal," Associated Press, November 12, 2011, http://townhall.com/news/politics-elections/2011/11/11/obama_urges_debt_panel_to _reach_for_deal.

36. Eli Lake, "Shite House Pressure for a Donor?" *Daily Beast*, September 15, 2011, http:// www.thedailybeast.com/articles/2011/09/15/lightsquared-did-white-house-pressure -general-shelton-to-help-donor.html.

37. "Philip Falcone," *Forbes*, http://www.forbes.com/profile/philip-falcone/.

38. Eli Lake, "White House's Testimony 'Guidance,'" *Daily Beast*, September 19, 2011, http://www.thedailybeast.com/articles/2011/09/19/lightsquared-second-witness-rejects -white-house-testimony-guidance.html.

39. Michelle Malkin, "LightSquared: Obama's Dangerous Broadband Boondoggle," *Townhall.com*, September 21, 2011, http://townhall.com/columnists/michellemalkin/ 2011/09/21/lightsquared_obamas_dangerous_broadband_boondoggle/page/full/.

40. Rachel Leven, "LightSquared doubles size of its lobbying team in 2011," *Hill*, September 27, 2011, http://thehill.com/business-a-lobbying/184031-lightsquared-doubles -size-of-its-lobbying-team-in-2011.

41. Timothy P. Carney, "Soros turns up in Obama's LightSquared imbroglio," *Washington Examiner*, September 13, 2006, http://campaign2012.washingtonexaminer.com/ article/soros-turns-obamas-lightsquared-imbroglio.

42. "Soros Surfaces on the Edge of White House Controversy Involving LightSquared," *FoxNews.com*, September 23, 2011, http://www.foxnews.com/politics/2011/09/23/ soros-surfaces-on-edge-white-house-controversy/.

43. Ibid.

44. Rowan Scarborough, "Lehman rocks Navy with complaints about political correctness," *Washington Times*, September 18, 2011, http://www.washingtontimes.com/news/2011/sep/18/lehman-rocks-navy-complaints-about-political-corre/.

45. Zuhdi Jasser, "The Islamist Threat Inside Our Military," *Wall Street Journal*, August 18, 2011, http://online.wsj.com/article/SB10001424053111904006104576500593598420746.html.

46. Frank Gaffney, "A smoking gun," *Washington Times*, June 28, 2011, http://www.washingtontimes.com/news/2011/jun/28/a-smoking-gun/.

47. Laura Meckler, "Obama Names Openly Gay Veteran to West Point Advisory Board," *Wall Street Journal*, July 5, 2011, http://blogs.wsj.com/washwire/2011/07/05/obama-names-openly-gay-veteran-to-west-point-advisory-board/.

48. Daniel Luzer, "Obama Appoints Lesbian Veteran to West Point Board," *Washington Monthly*, July 7, 2011, http://www.washingtonmonthly.com/college_guide/blog/president_appoints_lesbian_vet.php.

49. "Gay Graduate Appointed to West Point's Board," *Army Times*, July 5, 2011, http://www.armytimes.com/news/2011/07/ap-west-point-brenda-sue-fulton-board-of-visitors-070511/.

50. Lolita C. Baldor and Donna Cassata, "Military money on chopping block in austere time," Associated Press, August 4, 2011, http://www.seattlepi.com/news/article/Military-money-on-chopping-block-in-austere-time-1722001.php.

51. Niccolo Machiavelli, "Report: Obama Administration Bows To Chinese Pressure On Taiwan Arms Sales; Will Cost U.S. 16,000 Jobs," August 15, 2011, http://bigpeace.com/nmachiavelli/2011/08/15/report-obama-administration-bows-to-chinese-pressure-on-taiwan-arms-sales-will-cost-u-s-16000-jobs/.

Chapter 8. Tyranny of Treating Our Friends Like Enemies and Our Enemies Like Friends

1. "Obama rebounds in polls as economic crisis bites," Agence France Presse, September 18, 2008, http://afp.google.com/article/ALeqM5grzIRe03CerX8EC_7WHqwavChcMg.

2. "Sharansky's Insights on Egypt's Future," Sceptical Market Observer, February 5, 2011, http://scepticalmarketobserver.blogspot.com/2011/02/sharanskys-insights-on-egypts-future.html.

3. Michael Slackman, "Syrian Troops Open Fire on Protesters," Real Clear Politics, March 26, 2011, http://www.realclearpolitics.com/2011/03/26/syrian_troops_open_fire_on_protesters_252732.html.

4. Glenn Kessler, "Hillary Clinton's uncredible statement on Syria," *Washington Post*, April 4, 2011, http://www.washingtonpost.com/blogs/fact-checker/post/hillary-clintons-uncredible-statement-on-syria/2011/04/01/AFWPEYaC_blog.html.

5. Richard Falk, "Welcoming the Tunisian Revolution: Hopes and Fears," *Foreign Policy Journal*, January 24, 2011, http://www.foreignpolicyjournal.com/2011/01/24/welcoming-the-tunisian-revolution-hopes-and-fears/.

6. Patrick Poole, "New Law Cuts Ties Between FBI and Terror-Tied Groups," PJ Media,

November 22, 2011, http://pjmedia.com/blog/pjm-exclusive-new-law-cuts-ties-between
-fbi-and-terror-tied-groups/.

7. "Obama says Mubarak must begin 'transition' now," Jihad Watch, February 1, 2011,
    http://www.jihadwatch.org/2011/02/obama-says-mubarak-must-go-now.html.

8. "Obama Administration Corrects Clapper's Claim Muslim Brotherhood is 'Secu-
    lar,' " Fox News, February 10, 2011, http://www.foxnews.com/politics/2011/02/10/
    administration-corrects-dni-clapper-claim-muslim-brotherhood-secular/.

9. Leland Vittert, "Egyptian Businessman Fears Muslim Brotherhood Takeover," Fox
    News, February 8, 2011, http://www.foxnews.com/world/2011/02/08/leading
    -egyptian-businessman-fears-muslim-brotherhood/.

10. "Egypt's Christians, attacked by army, may flee country," Baptist Press, October 11,
    2011, http://www.bpnews.net/bpnews.asp?id=36312.

11. Keith Koffler, "Obama Responds to Egyptian Military's Massacre of Christians . . .
    By Asking Christians to Show Restraint," MidnightWatcher's Blogspot, October 13,
    2011, http://midnightwatcher.wordpress.com/2011/10/13/obama-responds-to-egyptian
    -military%E2%80%99s-massacre-of-christians-by-asking-christians-to-show-restraint/.

12. Arnaud De Borchgrave, "Lawlessness Rampant in Egypt," NewsMax, March 11, 2011,
    http://www.newsmax.com/deBorchgrave/egypt-hosni-mubarak-coptic/2011/03/11/
    id/389139.

13. Jonah Goldberg, "Where feminists need to fight," Townhall.com, March 30, 2011,
    http://www.nypost.com/p/news/opinion/opedcolumnists/where_feminists_need_to
    _fight_uloZpncrMT5NHoUrduQOnM.

14. Ian Birrell, "Devastating secret files reveal Labour lies over Gaddafi: Dictator warned
    of holy war if Lockerbie bomber Megrahi died in Scotland," MailOnline, Septem-
    ber 4, 2011, http://www.dailymail.co.uk/news/article-2033460/Secret-files-Labour-lied
    -Gaddafi—warned-holy-war-Megrahi-died-Scotland.html.

15. Jason Allardyce and Tony Allen-Mills, "White House backed release of Locker-
    bie bomber Abdel Baset al-Megrahi," Australian, July 26, 2010, http://www.the
    australian.com.au/news/world/white-house-backed-release-of-lockerbie-bomber-abdel
    -baset-al-megrahi/story-e6frg6so-1225896741041.

16. Kerry Picket, "U.S. recognized Libyan government working with Muslim Brother-
    hood?" Washington Times, August 22, 2011, http://www.washingtontimes.com/blog/
    watercooler/2011/aug/22/picket-us-recognized-libyan-government-working-mus/.

17. "How the Taliban Lost Its Swagger," Newsweek, February 27, 2011, http://www
    .thedailybeast.com/newsweek/2011/02/27/how-the-taliban-lost-its-swagger.html.

18. Elizabeth MacDonald, "How Much Did the U.S. Give Pakistan," Fox Business, May
    11, 2011, http://www.foxbusiness.com/markets/2011/05/11/did-pakistan/.

19. Asif Ali Zardari, "Pakistan did its part," Washington Post, May 2, 2011, http://www
    .washingtonpost.com/opinions/pakistan-did-its-part/2011/05/02/AFHxmybF_story
    .html.

20. Victor Volsky, "Obama's Courage Vastly Overrated in Pakistan Raid," American
    Thinker, May 6, 2011, http://www.americanthinker.com/2011/05/obamas_courage
    _vastly_overrate.html.

21. "Cuba: U.S. aid worker Alan Gross's trial ends," BBC Mobile, March 5, 2011, http://www
    .bbc.co.uk/news/world-latin-america-12657855.

22. Luisita Lopez Torregrossa, "Trial of U.S. Aid Contractor Alan Gross Ends in Havana," *Politics Daily*, March 6, 2011, http://www.politicsdaily.com/2011/03/06/trial -of-u-s-aid-contractor-alan-gross-ends-in-havana/.

23. Randal C. Archibold, "Cuba Gives 15-Year Prison Term to American," *New York Times*, March 12, 2011, http://www.nytimes.com/2011/03/13/world/americas/13cuba .html.

24. Hillary Busis, "Obama's Joke About Going to Visit Hugo Chavez Doesn't Go Over Very Well," Mediaite, November 22, 2010, http://www.mediaite.com/online/obamas -joke-about-going-to-visit-hugo-chavez-doesnt-go-over-very-well/.

25. Antonio Sosa, "Chavez responds to President Obama's joke about visiting Venezuela, says he would 'embrace' president and 'eat socialist arepas' with him," *Daily Caller*, November 22, 2010, http://dailycaller.com/2010/11/22/chavez-responds-to-president -obamas-joke-about-visiting-venezuela-says-he-would-embrace-president-and-eat-socialst -arepas-with-him/.

26. Busis, "Obama's Joke About Going to Visit Hugo Chavez Doesn't Go Over Very Well."

27. Jeremy Morgan, "Chavez Government Closes Venezuela Opposition Newspaper & Magazine," *Latin American Herald Tribune*, http://www.laht.com/article.asp?ArticleId =347744&CategoryId=10717.

28. "Hugo Chavez closes 34 Venezuelan radio stations," *Telegraph*, August 2, 2009, http:// www.telegraph.co.uk/news/worldnews/southamerica/venezuela/5961183/Hugo-Chavez -closes-34-Venezuelan-radio-stations.html.

29. Kevin Sullivan, "Chavez Tightening Grip on Judges, Critics Charge," *Washington Post*, June 20, 2004, http://www.washingtonpost.com/wp-dyn/articles/A54913-2004 Jun19.html.

30. Richard Miniter, "Obama, Chavez And The American Shareholder," *Forbes.com*, January 12, 2011, http://www.forbes.com/2011/01/12/hugo-chavez-expropriation-owens -illinois-opinions-contributors-richard-miniter.html.

31. "Venezuela faces growing load of arbitration cases," Associated Press, October 10, 2011, http://my.news.yahoo.com/venezuela-faces-growing-load-arbitration-cases-131818739 .html.

32. Miniter, "Obama, Chavez And The American Shareholder."

33. "Expropriated Companies in Venezuela Get Little to No Compensation," WNVz Blog, http://whatsnextvenezuela.com/expropriation/expropriated-companies-in-venezuela -get-little-to-no-compensation/.

34. Miniter, "Obama, Chavez And The American Shareholder."

35. "Food fight: How to destroy an industry," *Economist*, June 10, 2010, http://www.econ omist.com/node/16326418.

36. Rachel Jones, "Chavez Gives Himself a Big Christmas Gift," *Time*, December 29, 2010, http://www.time.com/time/world/article/0,8599,2039867,00.html.

37. "Ros-Lehtinen Calls on OAS to Stand Up to Chavez's Ongoing Power Grab," United States House, Committee on Foreign Affairs, November 22, 2010, http://foreign affairs.house.gov/press_display.asp?id=1671.

38. Roger Noriega, "Hugo Chavez's Big Lie," *FoxNews.com*, November 8, 2011, http:// www.foxnews.com/opinion/2011/11/08/hugo-chavezs-big-lie/.

39. Jordan Sekulow, "Obama's Schizophrenic Israel Policy," *Human Events*, May 25, 2011, http://www.humanevents.com/article.php?id=43711.

40. Ibid.

41. " 'We can't go back': Israeli PM rejects 1967 border proposal," *MSNBC.com*, May 20, 2011, http://www.msnbc.msn.com/id/43106082/ns/politics-white_house/t/we-cant -go-back-israeli-pm-rejects-border-proposal/.

42. Barry Rubin, "New Phase in War as Terrorists Cross Egypt-Israel Border, 7 Israelis Dead," Stand With Us, August 19, 2011, http://www.standwithus.com/app/inews/view _n.asp?ID=1967.

43. "Attack on Egypt embassy marks beginning of Israel's end, Iranian officials say," *Haaretz*, September 17, 2011, http://www.haaretz.com/news/diplomacy-defense/attack -on-egypt-embassy-marks-beginning-of-israel-s-end-iranian-officials-say-1.385031.

44. Patrick Goodenough, "U.S. and 'Quartet' Partners Criticize Israel but Sidestep Palestinians' Statehood Bid at U.N.," *News America Daily*, August 17, 2011, http://www .newsamericadaily.com/u-s-and-%E2%80%98quartet%E2%80%99-partners-criticize -israel-but-sidestep-palestinians%E2%80%99-statehood-bid-at-u-n/.

45. Itamar Marcus and Nan Jacques Zilberdik, "Mother of 4 terrorist murderers chosen by the PA to launch statehood campaign," Palestinian Media Watch, September 18, 2011, http://www.pmw.org.il/main.aspx?fi=157&doc_id=5656.

46. "Attack on Egypt embassy marks beginning of Israel's end, Iranian officials say," *Haaretz*, September 17, 2011, http://www.haaretz.com/news/diplomacy-defense/attack -on-egypt-embassy-marks-beginning-of-israel-s-end-iranian-officials-say-1.385031.

47. Eldad Tzioni, "White House vetoes UNSC vote—and then throws Israel under the bus," NewsRealBlog, February 21, 2011, http://www.newsrealblog.com/2011/02/21/ white-house-vetoes-unsc-vote-and-then-throws-israel-under-the-bus/.

48. Daniel Halper, "Obama Fundraiser and Ambassador Blames Israel for Anti-Semitism (Updated)," *The Weekly Standard*, December 3, 2011, http://www.weeklystandard .com/blogs/obama-fundraiser-and-ambassador-blames-israel-anti-semitism_610946 .html.

49. Chris Queen, "Not Invited to the Wedding: Why Great Britain Doesn't Like Barack Obama," NewsRealBlog, May 2, 2011, http://www.newsrealblog.com/2011/05/02/ not-invited-to-the-wedding-why-great-britain-doesnt-like-barack-obama/.

50. Ibid.

51. Ginger Thompson, "U.S. Suspends $30 million to Honduras," *New York Times*, September 30, 2009, http://www.nytimes.com/2009/09/04/world/americas/04honduras .html.

52. Rory Carroll, "Honduras elects Porfirio Lobo as new president," *Guardian*, November 30, 2009, http://www.guardian.co.uk/world/2009/nov/30/honduras-lobo -president.

53. Aaron Klein, "Largest U.S. union behind Mideast riots?" *WorldNetDaily*, August 8, 2011, http://www.wnd.com/?pageId=331509.

54. Aaron Klein, "U.S. 'held secret meeting with Muslim Brotherhood,' " *WorldNetDaily*, February 1, 2011, http://www.wnd.com/?pageId=258405.

55. Zvi Bar'el and Avi Issachiroff, "Obama met Muslim Brotherhood members in U.S.,"

Haaretz, April 6, 2009, http://www.haaretz.com/news/obama-met-muslim-brother hood-members-in-u-s-1.277306.

56. "Palestinian Rockets Target Israeli School Bus," CBN News, April 7, 2011, http://www.cbn.com/cbnnews/insideisrael/2011/April/Palestinians-Target-Israeli-School -Bus/.

57. Nasser Karimi, "Iran's Ahmadinejad: No Place for Israel in Region," Associated Press, August 26, 2011, http://www.cnsnews.com/news/article/irans-ahmadinejad-no -place-israel-region.

58. Bruce McQuain, "Meanwhile, In the Middle East—Saudi/US Strains Appear," Questions and Observations: Free Markets, Free People, March 15, 2011, http://www.qando.net/?p=10512.

59. Maggie Michael and Diaa Hadid, "Israel Embassy Attacked: Prime Minister Benjamin Netanyahu Condemns Violence in Egypt," *Huffington Post*, September 10, 2011, http://www.huffingtonpost.com/2011/09/10/israel-embassy-attacked-egypt_n_956627.html.

60. Kerry Picket, "Obama Administration: Room for Muslim Brotherhood in reformed Egyptian government," *Washington Times*, February 1, 2011, http://www3.washing tontimes.com/blog/watercooler/2011/feb/1/obama-administration-room-muslim-brother hood-refor/.

61. Douglas Hamilton, "Israel shocked by Obama's 'betrayal' of Mubarak," Reuters, January 31, 2011, http://www.reuters.com/article/2011/01/31/us-egypt-israel-usa-idUSTRE70 U53720110131.

Chapter 9. Tyranny of Green Energy

1. "Henry Juskiewicz," *Bloomberg Businessweek*, http://investing.businessweek.com/busi nessweek/research/stocks/private/person.asp?personId=4339939&privcapId=28990& previousCapId=255397&previousTitle=BEST%20BUY%20CO%20INC.

2. James R. Hagerty and Kris Maher, "Gibson Guitar Wails on Federal Raid Over Wood," *Wall Street Journal*, September 1, 2011, http://online.wsj.com/article/SB10001 4240531119038959045765429420278593286.html.

3. "Gibson Guitar Corp. Responds to Federal Raid," *Gibson.com*, August 25, 2011, http://www.gibson.com/en-us/Lifestyle/News/gibson-0825-2011/.

4. "Guitar Frets: Environmental Enforcement Leaves Musicians in Fear," *Wall Street Journal*, August 26, 2011, http://online.wsj.com/article/SB1000142405311190478740 4576530520471223268.html.

5. Nancy DeWolf Smith, "The Government and the Guitar Man," *Wall Street Journal*, November 12, 2011, http://online.wsj.com/article/SB10001424052970203554104576 655273915372748.html?mod=WSJ_Opinion_LEADTop)

6. Kate Hicks, "Solyn-drama: Obama Warned Not to Visit Failed Solar Panel Plant," *Townhall.com*, October 4, 2011, http://townhall.com/tipsheet/katehicks/2011/10/03/ solyn-drama_obama_warned_not_to_visit_failed_solar_panel_plant.

7. Jim McElhatton, "Solyndra put off word of layoffs until after election," *Washington Times*, November 15, 2011, http://www.washingtontimes.com/news/2011/nov/15/ solyndra-put-off-word-of-layoffs-until-after-elect/.

8. Michael Barone, "Obama Tainted by Loan Guarantees to Solar Firms," *Washington Examiner*, September 14, 2011, http://www.aei.org/article/104134?gclid=CKLj0_Warqs CFaUCQAod7WYYJQ.

9. Carolyn Lochhead, "Energy stimulus program plagued by problems," *SFGate.com*, November 3, 2011, http://articles.sfgate.com/2011-11-03/news/30357933_1_stimulus -funds-stimulus-program-stimulus-money.

10. Carol Leonnig and Joe Stephens, "Solyndra employees: Company suffered from mismanagement, heavy spending," *Washington Post*, September 21, 2011, http://www .washingtonpost.com/politics/solyndra-employees-company-suffered-from-mismanage ment-heavy-spending/2011/09/20/gIQAMHC3lK_story.html.

11. Deborah Solomon and Ryan Tracy, "Firm Denied Troubles in Spring," *Wall Street Journal*, September 17, 2011, http://online.wsj.com/article/SB1000142405311190392 7204576574912734624144.html.

12. Jim McElhatton, "Bankruptcy scenario played out before Solyndra collapse," *Washington Times*, September 22, 2011, http://www.washingtontimes.com/news/2011/ sep/22/bankruptcy-scenario-played-out-before-solyndra-col/.

13. Guy Benson, "Disgrace: Solyndra Execs Awarded Themselves Huge Bonuses as Company Crumbled," *Townhall.com*, November 4, 2011, http://townhall.com/tipsheet/ guybenson/2011/11/04/disgrace_solyndra_execs_awarded_themselves_huge_bonuses _as_company_crumbled.

14. John Hinderaker, "Solyndra Execs Leave With Cash. Yours," PowerLine, November 2, 2011, http://www.powerlineblog.com/achives/2011/11/solyndra-execs-leave-with -cash-yours.php.

15. Guy Benson, "BOMBSHELL: Emails Reveal Direct White House Tie to Solyndra Scandal," *Townhall.com*, September 14, 2011, http://townhall.com/tipsheet/guy benson/2011/09/14/bombshell_emails_reveal_direct_white_house_tie_to_solyndra _scandal.

16. "PUBLIC LAW 109-58—AUG. 8, 2005, ENERGY POLICY ACT OF 2005," p. 499, http://doi.net/iepa/EnergyPolicyActof2005.pdf.

17. Ibid., p. 526.

18. Andrew Stiles, "Solyndra Loan: DOE Memo Defends Decision to Put Private Investors Ahead of Taxpayers," *National Review Online*, October 14, 2011, http://www.nation alreview.com/corner/280161/solyndra-loan-doe-defends-decision-put-private-investors -ahead-taxpayers-andrew-stiles.

19. "The Solyndra Economy," *Wall Street Journal*, October 11, 2011, http://online.wsj .com/article/SB10001424052970204524604576610972882349418.html?mod=WSJ _Opinion_LEADTop.

20. Bruce Krasting, "Throw the Bums Out," *Bruce Krasting: My Take on Financial Events*, December 18, 2011, http://brucekrasting.blogspot.com/2011/12/throw-bums-out.html.

21. "House Panel Votes to Subpoena White House for Solyndra Records," *FoxNews.com*, November 3, 2011, http://www.foxnews.com/politics/2011/11/02/obama-administration -considered-bailout-for-solyndra-before-bankruptcy/.

22. Jim McElhatton, "FBI-targeted solar company Solyndra to put shine on law firm's wallet," *Washington Times*, September 19, 2011, http://www.washingtontimes.com/news/ 2011/sep/19/fbi-targeted-solar-company-to-put-shine-on-law-fir/?page=all.

23. Martin Gould, "$737 million Green Loan to Pelosi Kin Fuels Outrage," *NewsMax.com*, September 30, 2011, http://www.newsmax.com/Headline/New-Green-Loan-Obama/2011/09/30/id/412909.

24. "Greasing the solar skids," *New York Post*, November 18, 2011, http://www.nypost.com/p/news/opinion/editorials/greasing_the_solar_skids_yf8pKoJEcktj7gUSFOWurO.

25. "Another Energy Company Goes Bankrupt, $39 million Borrowed From Taxpayers," *FoxNews.com*, October 31, 2011, http://www.foxnews.com/politics/2011/10/31/another-energy-company-goes-bankrupt-3-million-borrowed-from-taxpayers/.

26. "Beacon Power's Catastrophic Flywheel Failures," Energy Storage Blog, October 26, 2011, http://www.energystorageblog.com/2011/10/26/beacon_powers_catastrophic_flywheel_failures/.

27. "Is Ener1 Next for Bankruptcy?" Energy Storage Blog, November 1, 2011, http://www.energystorageblog.com/2011/11/01/is-ener1-next-for-bankruptcy/.

28. "Solyndra Not Sole Firm to Hit Rock Bottom Despite Stimulus Funding," Fox News, September 15, 2011, http://www.foxnews.com/politics/2011/09/15/despite-stimulus-funding-solyndra-and-4-other-companies-have-hit-rock-bottom/.

29. John Nichols and Paul Driessen, "Delaware's very own Solyndra?" *Washington Times*, September 23, 2011, http://www.washingtontimes.com/news/2011/sep/23/delawares-very-own-solyndra/.

30. Darius Dixon, "DOE IG: 100+ stimulus-related criminal probes," *Politico*, November 2, 2011, http://www.politico.com/news/stories/1111/67444.html.

31. Neil King, Neil and Stephen Power, "Times Tough for Energy Overhaul," *Wall Street Journal*, December 12, 2008, http://online.wsj.com/article/SB122904040307499791.html.

32. "Steven Chu Biography," Republican National Convention blog, December 15, 2008, http://rncnyc2004.blogspot.com/2008/12/steven-chu-biography.html.

33. "Price Charts," GasBuddy.com, http://gasbuddy.com/gb_retail_price_chart.aspx_.

34. "Obama says he wants oil producers to boost output," *Bloomberg Businessweek*, April 26, 2011, http://www.businessweek.com/ap/financialnews/D9MRJH8O0.htm.

35. Jim Kuhnhenn, "Obama says he wants oil producers to boost output," *Bloomberg Businessweek*, April 27, 2011, http://www.businessweek.com/ap/financialnews/D9MRQU9G2.htm.

36. Mark Knoller, "National Debt Up $3 Trillion on Obama's Watch," Political Hotsheet, October 18, 2011, http://www.cbsnews.com/8301-503544_162-20019931-503544.html.

37. "Team to probe oil market fraud, manipulation: Obama," Reuters, April 22, 2011, http://www.reuters.com/article/2011/04/22/us-obama-oil-idUSTRE73K6WW20110422.

38. Mark J. Perry, "In Q1 Exxon Paid Almost $1 million per Hour in Income Taxes," Seeking Alpha, April 29, 2011, http://seekingalpha.com/article/266554-in-q1-exxon-paid-almost-1-million-per-hour-in-income-taxes.

39. Ibid.

40. Opelka, Mike, "Exxon Earns Huge Profits, But Also Pays Huge Taxes," *Blaze*, April 29, 2011, http://www.theblaze.com/stories/exxon-earns-huge-profits-but-also-pays-huge-taxes/.

41. "Exxon Mobil Corp Earnings Conference Call (Q1 2011)," Yahoo! Finance, April 28, 2011, http://biz.yahoo.com/cc/0/121530.html.

42. Mark Perry, "Q1: Exxon Paid Almost $1M Per Hour In Income Taxes; Its Effective Tax Rate Was 42.3%," Daily Markets, April 29, 2011, http://www.dailymarkets .com/economy/2011/04/28/qi-exxon-paid-almost-1m-per-hour-in-income-taxes-and-its -effective-tax-rate-was-42-3/.

43. Lucian Pugliarese, "The Keystone Debacle," *Wall Street Journal*, November 16, 2011, http://online.wsj.com/article/SB10001424052970204190504577037754000084544 .html?mod=WSJ_Opinion_LEFTTopOpinion.

44. Christie Parsons and Paul Richter, "Keystone XL pipeline decision delayed until after 2012 election," *Los Angeles Times*, November 11, 2011, http://www.latimes.com/news/ nationworld/nation/la-na-keystone-20111111,0,1732793.story.

45. Pugliarese, "The Keystone Debacle."

46. José De Córdoba, "Reports of Chavez's Illness Cloud Campaign," *Wall Street Journal*, November 19, 2011, http://online.wsj.com/article/SB10001424052970204517204577 046464037810838.html?mod=WSJ_World_MIDDLENews.

47. Zachary Goldfarb, "A Regulator Heeds Lessons From the Past," *Washington Post*, July 24, 2009, http://www.washingtonpost.com/wp-dyn/content/article/2009/07/23/AR 2009072303609.html.

48. "The Talented Mr. Gensler," *The Wall Street Journal*, December 12, 2011, http://online .wsj.com/article/SB10001424052970204903804577080480063317856.html.

49. Jeff Mason, "Symbolism and Sympathy: Obama's risky gasoline strategy," Reuters, May 1, 2011, http://www.reuters.com/article/2011/05/01/us-obama-gasoline-idUSTR E7401G920110501.

50. "Obama Eyes Speculators for Rising Gas Prices, as Other Factors Play Role," Fox News, April 22, 2011, http://www.foxnews.com/politics/2011/04/22/obama-form-task-force -tackle-rising-gas-prices/.

51. Andrew Restuccia, " 'I'm a Shell Oil creation,' says EPA chief," *Hill*, April 26, 2011, http://thehill.com/blogs/e2-wire/677-e2-wire/157741-jackson-im-a-shell-oil-creation.

52. "EPA Head Lisa Perez Jackson's Radical Brand of Environmental Justice," *Liberty Journal*, February 18, 2010, http://thelibertyjournal.com/2010/02/18/epa-head-lisa -perez-jacksons-radical-brand-of-environmental-justice/.

53. Martin Gould, "Shell Oil Lambastes Odd EPA Denial on Drilling," *NewsMax.com*, April 26, 2011, http://www.newsmax.com/US/Shell-oil-drilling-EPA/2011/04/26/id/ 394138.

54. Stephen Power, "Senate Halts Effort to Cap CO_2 Emissions," *Wall Street Journal*, July 23, 2010, http://online.wsj.com/article/SB10001424052748703467304575538337360 0358634.html.

55. "A Majority of the House and Senate Vote to Limit EPA's Regulations," Institute for Energy Research, April 13, 2011, http://www.instituteforenergyresearch .org/2011/04/13/a-majority-of-the-house-and-senate-vote-to-limit-epa%E2%80%99s -regulations/.

56. "Clean Air Act Timeline," http://www.edf.org/documents/2695_cleanairact.htm.

57. "Clean Air Act Timeline," Clear the Air, http://www.edf.org/page.cfm?tagID=60844.

58. "The EPA's Reliability Cover-Up," *Wall Street Journal*, November 14, 2011, http://online .wsj.com/article/SB10001424052970204358004577030110213488278.html.

59. Ibid.

60. "Global Warming: Why Can't the Mainstream Press Get Even Basic Facts Right?" National Center for Public Policy Research, March 22, 2004, http://www.national center.org/TSR032204.html.

61. "IPCC Official: Climate Policy Is Redistributing The World's Wealth," US Message Board, November 19, 2010, http://www.usmessageboard.com/economy/142699-we -redistribute-worlds-wealth-by-climate-policy.html.

62. Charles Biggs, "Thousands of scientists sign petition against global warming," *Tulsa Beacon*, June 5, 2008, http://www.tulsabeacon.com/?p=462.

63. "CERN: The Sun Causes Global Warming," *EU Times*, September 3, 2011, http:// www.eutimes.net/2011/09/cern-the-sun-causes-global-warming/.

64. Terry Hurlbut, "Hadley CRU hacked with release of hundreds of docs and emails," *Examiner.com*, November 19, 2009, http://www.examiner.com/essex-county-conserv ative-in-newark/hadley-cru-hacked-with-release-of-hundreds-of-docs-and-emails.

65. "The Rules of the Game," http://www.futerra.co.uk/downloads/RulesOfTheGame .pdf.

66. Jurrian Maessen, "Unearthed Files Include 'Rules' for Mass Mind Control Campaign," Infowars, November 25, 2009, http://www.infowars.com/unearthed-files-include -rules-for-mass-mind-control-campaign/.

67. "A Majority of the House and Senate Vote to Limit EPA's Regulations," Insti- tute for Energy Research, April 13, 2011, http://www.instituteforenergyresearch .org/2011/04/13/a-majority-of-the-house-and-senate-vote-to-limit-epa%E2%80%99s regulations/.

68. Ibid.

69. Mark Clayton, "Global warming: Congress set to decide if EPA can regulate green- house gases," *Christian Science Monitor*, April 6, 2011, http://www.csmonitor.com/ USA/Politics/2011/0406/Global-warming-Congress-set-to-decide-if-EPA-can-regulate -greenhouse-gases.

70. Sabrina Valle, "Losing Forests to Fuel Cars," *Washington Post*, July 31, 2007, http:// www.washingtonpost.com/wp-dyn/content/article/2007/07/30/AR2007073001484 .html.

71. Glenn Hurowitz, "Soros, Goldman Sachs financing destruction of Brazilian forests," Grist, August 2, 2007, http://www.grist.org/article/george-soros-vs-the-planet.

72. Rodrigo Orihuela, "Soros-Backed Adecoagro Raises $314 million in IPO," *Bloomberg Businessweek*, January 28, 2011, http://www.businessweek.com/news/2011-01-28/ soros-backed-adecoagro-raises-314-million-in-ipo.html.

73. "Senators Feinstein, Coburn Introduce Bill to Eliminate Ethanol Subsidy, Tariff," United States Senator Dianne Feinstein, May 3, 2011, http://feinstein.senate.gov/ public/index.cfm?FuseAction=NewsRoom.PressReleases&ContentRecord_id=b7bfdc b4-5056-8059-766d-4ea48cce735d&IsPrint=true.

74. Sabrina Valle, "Losing Forests to Fuel Cars," *Washington Post*, July 31, 2007, http://www .washingtonpost.com/wp-dyn/content/article/2007/07/30/AR2007073001484.html.

75. "Soros hedge fund invests $811 million to buy Petrobras stake," Sweetness & Light, August 15, 2008, http://sweetness-light.com/archive/soros-invests-811m-in-brazilian-oil.

76. Dominic Rushe, "George Soros to close hedge fund management group to outside investors," *Guardian*, July 26, 2011, http://www.guardian.co.uk/business/2011/jul/26/george-soros-to-close-hedge-fund.

77. Ed Lasky, "Keep an eye on InterOil & George Soros," *Pittsburgh Tribune-Review*, April 3, 2011, http://www.pittsburghlive.com/x/pittsburghtrib/opinion/s_730320.html.

78. Amanda Carey, "Report says U.S. has largest fossil fuel reserves in world," *Daily Caller*, March 10, 2011, http://dailycaller.com/2011/03/10/new-report-says-u-s-has-largest-fossil-fuel-reserves-in-world/.

79. Ben German, "It's George Soros versus Exxon in fight over oil payment disclosures," *Hill*, March 1, 2011, http://thehill.com/blogs/e2-wire/677-e2-wire/146749-its-george-soros-against-exxon-on-oil-payments-disclosure.

80. Ed Lasky, "George Soros Handicapping American Energy," *American Thinker*, March 2, 2011, http://www.americanthinker.com/2011/03/george_soros_handicapping_amer.html.

81. Giuseppe Marconi, "There will be no new refineries," *Oil-Price.net*, July 23, 2008, http://www.oil-price.net/en/articles/oil-refineries.php.

82. Ben Smith, "Hochberg to head Ex-Im," *Politico*, January 9, 2009, http://www.politico.com/blogs/bensmith/0109/Hochberg_to_head_ExIm.html?showall.

83. "The New School," http://the-new-school.co.tv/#Founding.

84. Douglas Kellner, "Herbert Marcuse," *Illuminations*, http://www.uta.edu/huma/illuminations/kell12.htm.

85. "PEBO to Name Fred Hochbert to Head Export-Import Bank," ABC News, January 9, 2009, http://blogs.abcnews.com/politicalpunch/2009/01/pebo-to-name-fr.html.

86. "Export-Import Bank of the United States," Fed Up USA, http://fedupusa.wordpress.com/export-import-bank-of-the-united-states/.

87. "Milano Reunion 2007," http://www.newschool.edu/milano/alumninewslettersp07.pdf.

88. Stephen Dinan, "Obama climate czar has socialist ties," *Washington Times*, January 12, 2009, http://www.washingtontimes.com/news/2009/jan/12/obama-climate-czar-has-socialist-ties/.

89. David Limbaugh, "Obama's Scandalous War Against Domestic Oil," *Townhall.com*, May 6, 2011, http://townhall.com/columnists/davidlimbaugh/2011/05/06/obamas_scandalous_war_against_domestic_oil.

90. Ibid.

91. Pete Kasperowicz, "Obama administration floats draft plan to tax cars by the mile," *Hill*, May 5, 2011, http://thehill.com/blogs/floor-action/house/159397-obama-floats-plan-to-tax-cars-by-the-mile.

92. George Soros, *The Age of Fallibility: Consequences of the War on Terror* (New York: PublicAffairs, 2006), Prologue, p. xvi.

Chapter 10. Tyranny of the Anti-Justice Department

1. "Iranian plot to kill Saudi ambassador thwarted, U.S. officials say," CNN Justice, October 11, 2011, http://articles.cnn.com/2011-10-11/justice/justice_iran-saudi-plot _1_informant-iranian-plot-saudi-arabia?_s=PM:JUSTICE.
2. "Journalist: Obama Lied to 9/11 Victims' Families About Terrorist Trials," News maxTV, May 17, 2011, http://www.youtube.com/watch?v=5pSLMCWaaD8.
3. Nathan Koppel, "Holder Shines Spotlight on CIA Interrogation Methods," *Wall Street Journal*, June 30, 2011, http://blogs.wsj.com/law/2011/06/30/eric-holder-shines -spotlight-on-cia-interrogation-methods/.
4. Byron York, "Holder defends DOJ 'kids' who represented Gitmo detainees," *Washington Examiner*, April 14, 2010, http://washingtonexaminer.com/blogs/beltway -confidential/holder-defends-doj-039kids039-who-represented-gitmo-detainees.
5. Thomas Joscelyn, "Meet the DOJ Lawyers Who Defended Terrorist Detainees," *Weekly Standard*, March 3, 2010, http://www.weeklystandard.com/blogs/meet-doj -lawyers-who-defended-terrorist-detainees.
6. "Journalist: Obama Lied to 9/11 Victims' Families About Terrorist Trials," News-maxTV, May 17, 2011, http://www.youtube.com/watch?v=5pSLMCWaaD8.
7. Steven Emerson, "Justice Dept. Keeping Islamic Bank Settlement Secret," *Jewish World Review*, April 29, 2011, http://www.jewishworldreview.com/0811/emerson 082911.php3.
8. "Obama Orders Launched Fast and Furious," YouTube.com, http://www.youtube .com/watch?v=-PNhYk9NuNc.
9. A. W. R. Hawkins, "Eric Holder Feigns Ignorance of Operation 'Fast and Furious' Now, But He Bragged of Overseeing Its Implementation in 2009," Andrew Breitbart's Big Government, July 8, 2011, http://biggovernment.com/awrhawkins/2011/07/08/ eric-holder-feigns-ignorance-of-operation-fast-and-furious-now-but-he-bragged-of-over seeing-its-implementation-in-2009/.
10. A. W. R. Hawkins, " 'Fast and Furious' Goes All the Way to the White House," *Human Events*, August 9, 2011, http://www.humanevents.com/article.php?id= 45398.
11. William La Jeunesse, "U.S. Government Used Taxpayer Funds to Buy, Sell Weapons During 'Fast and Furious,' Documents Show," Fox News, September 26, 2011, http:// www.foxnews.com/politics/2011/09/26/us-government-bought-and-sold-weapons -during-fast-and-furious-documents-show/.
12. United States Department of Justice, Office of the Inspector General, Evaluation and Inspections Division, "Review of ATF's Project Gunrunner," November 2010, p. v, http://www.justice.gov/oig/reports/ATF/e1101.pdf.
13. Doug Book, "Is Project Gunrunner Gun Control Chicago-Style?" *FloydReports.com*, June 20, 2011, http://www.exposeobama.com/2011/06/20/is-project-gunrunner-gun -control-chicago-style/.
14. James Walsh, " 'Fast and Furious' Weapons Used by Drug Cartels," *Newsmax.com*, June 27, 2011, http://www.newsmax.com/JamesWalsh/mexican-drug-cartels-fastand furious/2011/06/27/id/401528.

15. Michael A. Walsh, "Lethal Fiasco," *New York Post*, July 29, 2011, http://www.nypost .com/p/news/opinion/opedcolumnists/lethal_fiasco_rllfHYxZwgAHwNGmMUC aoL.

16. Walsh, " 'Fast and Furious' Weapons Used by Drug Cartels."

17. Walsh, "Lethal Fiasco."

18. Sari Horowitz, "A gunrunning sting gone fatally wrong," *Washington Post*, July 25, 2011, http://www.washingtonpost.com/investigations/us-anti-gunrunning-effort-turns -fatally-wrong/2011/07/14/gIQAH5d6YI_story.html.

19. "GOA Calls for House Investigation into 'Project Gunrunner' As Senator Reid Drags His Feet," Gun Owners of America, February 23, 2011, http://gunowners.org/project gunrunner.htm.

20. "Phoenix Special Agent John Dodson's Congressional testimony on Operation Fast & Furious," What the Folly?, June 22, 2011, http://www.whatthefolly.com/2011/06/22/ phoenix-atf-special-agent-john-dodsons-congressional-testimony-on-operation-fast -furious/.

21. Ibid.

22. "Bortac History and Overview," http://bortac.com/.

23. "Three indicted in killing of Border Patrol Agent Brian Terry," Sea to Shining Sea, May 7, 2011, http://seatoshiningsea.wordpress.com/2011/05/07/blog-post-three -indicted-in-killing-of-border-patrol-agent-brian-terry-1-suspect-in-custody-2-others -unnamed-and-at-large-feds-deny-any-wrongdoing-on-fast-and-furious/.

24. Mike Levine, "Three Men Charged In Murder Of Border Patrol Agent Brian Terry; Questions Over U.S. Program Linger," *FoxNews.com*, May 6, 2011, http://politics .blogs.foxnews.com/2011/05/06/three-men-charged-murder-border-patrol-agent-brian -terry-questions-over-us-program-linger.

25. Ibid.

26. Mark Krikorian, "Bringing a Bean Bag to a Gunfight," *National Review Online*, March 3, 2011, http://www.nationalreview.com/corner/261283/bringing-bean-bag -gunfight-mark-krikorian#.

27. Sharyl Attkisson, "Gunrunning scandal uncovered at ATF," CBS News, February 23, 2011, http://www.cbsnews.com/stories/2011/02/23/eveningnews/main20035609 .shtml.

28. Katie Pavlich, "Operation Fast and Furious: Coverup Continues With New Secret Recording," *Townhall.com*, September 19, 2011, http://tomdelay.townhall.com/tipsheet/ katiepavlich/2011/09/19/operation_fast_and_furious_coverup_continues_with_new _secret_recording.

29. William La Jeunesse, "America's Third War: Agent Brian Terry, A Policy of Silence?" *FoxNews.com*, February 22, 2011, http://www.foxnews.com/us/2011/02/22/ agent-brian-terry-policy-silence/#ixzz1EiWUs74d.

30. "Chairman Issa Subpoenas ATF for 'Project Gunrunner' Documents," Darryl Issa, April 1, 2011, http://issa.house.gov/index.php?option=com_content&view=article&id =726:chairman-issa-subpoenas-atf-for-project-gunrunner-documents&catid=63:2011 -press-release&Itemid=4.

31. "Letter From Attorney General Eric Holder to Charles Grassley," February 4, 2011,

http://grassley.senate.gov/about/upload/Judiciary-ATF-02-04-11-letter-from-DOJ-deny
-allegations.pdf.

32. Zach Lindsey, "Fast guns, furious Mexicans," *Mexico Weekly*, March 11, 2011, http://
newsessentials.wordpress.com/2011/04/02/fast-guns-furious-mexicans/.

33. "Three indicted in killing of Border Patrol Agent Brian Terry," Sea to Shining
Sea, May 7, 2011, http://seatoshiningsea.wordpress.com/2011/05/07/blog-post-three
-indicted-in-killing-of-border-patrol-agent-brian-terry-1-suspect-in-custody-2-others
-unnamed-and-at-large-feds-deny-any-wrongdoing-on-fast-and-furious/.

34. Sharyl Attkisson, "Agent: I was ordered to let U.S. guns into Mexico," CBS News,
March 3, 2011, http://www.cbsnews.com/stories/2011/03/03/eveningnews/main2003
9031.shtml.

35. "The Department of Justice's Operation Fast and Furious: Accounts of ATF Agents,"
Joint Staff Report Prepared for Rep. Darrell E. Issa and Senator Charles E. Grassley,
June 14, 2011, http://grassley.senate.gov/judiciary/upload/ATF-06-14-11-Joint-Issa
-Grassley-report-on-agent-findings.pdf.

36. Jerry Seper, " 'Fast & Furious': How botched operation spawned fatal results," *Washington Times*, October 17, 2011, http://www.washingtontimes.com/news/2011/oct/17/
fast-furious-how-a-botched-operation-spawned-fatal/?page=all.

37. "Report: Four ATF Agents Working on Controversial Operation 'Fast and Furious'
Tell Their Story," June 14, 2011, http://grassley.senate.gov/news/Article.cfm?custom
el_dataPageID_1502=35390.

38. M. Catharine Evans, "DOJ Letter Links 'Fast and Furious' To U.S. Crime Spree," Potter Williams Report, August 19, 2011, http://potterwilliamsreport.com/2011/08/18/
doj-letter-links-fast-and-furious-to-us-crime-spree.aspx.

39. William La Jeunesse, "Evidence Suggests Cover-Up in ATF Scandal, as More Guns
Appear at Crime Scenes," Fox News, September 2, 2011, http://www.foxnews.com/
politics/2011/09/02/demand-for-more-answers-in-fast-and-furious-scandal/.

40. "Letter From Senator Charles Grassley and Congressman Darryl Issa to U.S. Attorney General Eric Holder," July 5, 2011, http://oversight.house.gov/images/stories/
Letters/2011-07-05%20ceg-dei%20to%20ag.pdf.

41. "The Department of Justice's Operation Fast and Furious: Fueling Cartel Violence,"
Joint Staff Report, July 26. 2011, http://abcnews.go.com/images/Politics/FINAL%20
FINA.pdf.

42. Whitney Ksiazek, "Issa to DOJ: We've been Gamed," *FoxNews.com*, August 31, 2011,
http://politics.blogs.foxnews.com/2011/08/31/issa-doj-we-ve-been-gamed.

43. Michael Levine, " 'Fast & Furious'—An Insane Idea That Should Have Died At
Birth," Michael Levine Consulting and Investigations, July 27, 2011.

44. Ibid.

45. Katie Pavlich, "Obama Administration Targeting Border State Gun Shops . . . Again,"
Townhall.com, July 11, 2011, http://townhall.com/tipsheet/katiepavlich/2011/07/11/
obama_administration_targeting_border_state_gun_shopsagain.

46. Katie Pavlich, "Unbelievable: Operation Fast and Furious Agents . . . Promoted,"
Townhall.com, August 16, 2011, http://townhall.com/tipsheet/katiepavlich/2011/08/
16/unbelievable_operation_fast_and_furious_agentspromoted.

47. Michelle Malkin, " 'Furious' coverup?" *New York Post*, August 30, 2011, http://www
.nypost.com/p/news/opinion/opedcolumnists/furious_coverup_WkYocS6w6PStD71
TNxZRRM.

48. Edwin Mora, "Obama: Idea of Changing Immigration Laws on My Own 'Is Very
Tempting,' " *CNSNews.com*, July 26, 2011, http://www.cnsnews.com/news/article/
obama-idea-changing-immigration-laws-my.

49. Jim Hoft, "Figures. Obama Praises Illegal Aliens In Campaign Speech," Gateway Pundit, April 21, 2011, http://gatewaypundit.rightnetwork.com/2011/04/figures-obama
-praises-illegal-aliens-during-campaign-speech/.

50. Edwin Mora, "Labor Dept. Signs 'Partnerships' with Foreign Gov's to Protect Illegal
Workers in U.S.," *CNSNews.com*, August 31, 2011, http://cnsnews.com/news/article/
administration-partners-foreign-gov-s-pr.

51. Christopher Santarelli, "Former La Raza VP Cecilia Munoz Named White House
Domestic Policy Director," *The Blaze*, January 10, 2012, http://www.theblaze.com/
stories/former-la-raza-vp-cecilia-munoz-named-white-house-domestic-policy-director/.

52. Anna Schechter, "Born in the U.S.A.: Birth tourists get instant U.S. citizenship for
their newborns," Rock Center, October 28, 2011, http://rockcenter.msnbc.msn.com/
_news/2011/10/28/8511587-born-in-the-usa-birth-tourists-get-instant-us-citizenship-for
-their-newborns.

53. Alicia A. Caldwell, "Obama Administration Grants De Facto Amnesty to Many Illegal Immigrants," *CNSNews.com*, August 19, 2011, http://conservativeoutpost.com/
obama_administration_grants_de_facto_amnesty_many_illegal_immigrants.

54. Matt Cover, "HHS: Obamacare-Funded Health Centers for 'Migrants' Won't Check
Immigration Status," *CNSNews.com*, August 10, 2011, http://cnsnews.com/news/
article/hhs-obamacare-funded-health-centers-migrants-wont-check-immigration-status.

55. Katie Pavlich, "Obama Justice Department's Latest Target: Utah," Townhall.com,
November 23, 2011, http://townhall.com/tipsheet/katiepavlich/2011/11/23/obama
_justice_departments_latest_target_utah.

56. "Immunity from Justice," *Washington Times*, November 21, 2011, http://www.wash
ingtontimes.com/news/2011/nov/21/immunity-from-justice/?utm_source=RSS_Feed
&utm_medium=RSS.

57. Dave Gibson, "Illegal aliens received billions in tax credits last year," *Examiner.com*,
September 2, 2011, http://www.examiner.com/immigration-reform-in-national/illegal
-aliens-received-billions-federal-tax-credits-last-year.

58. Penny Starr, "Arizona Sheriff: 'Our Own Government Has Become Our Enemy,' "
CNSNews.com, August 1, 2010, http://cnsnews.com/news/article/arizona-sheriff-our
-own-government-has-become-our-enemy.

59. "Attacks on Border Patrol Agents 'Relatively Stable'—at More Than 1,000 Per Year,
Says DOJ," *CNSNews.com*, September 28, 2011, http://cnsnews.com/news/article/
attacks-border-patrol-agents-relatively-stable-more-1000-year-says-doj.

60. Jerry Seper, "U.S. border agent jailed for improper arrest of suspected drug smuggler,"
Washington Times, October 25, 2011, http://www.washingtontimes.com/news/2011/
oct/25/border-agent-jaile-arrest-teen-drug-smuggler/?page=1.

61. Edwin Mora, "Union President Gives Congress Emails Backing Testimony: ICE Or-

dered Agents Not to Arrest Illegals," *CNSNews.com*, November 8, 2011, http://www
.google.com/url?sa=t&rct=j&q=&esrc=s&source=web&cd=2&ved=0CCkQFjAB&
url=http%3A%2F%2Fwww.cnsnews.com%2Fnews%2Farticle%2Funion-president
-gives-congress-emails-backing-testimony-ice-ordered-agents-not-arrest&ei=wCTFTs
PhLYfg0QGtpvzsDg&usg=AFQjCNExXCAS3GuUM Ey4E5dD6nfBevL87Q.

62. Edwin Mora, "DOJ: 'Mexican-Based Trafficking Organizations Control Access to the
U.S.-Mexico Border,'" *CNSNews.com*, September 28, 2011, http://cnsnews.com/news/
article/doj-mexican-based-trafficking-organizations-control-access-us-mexico-border.

63. Jerry Seper, "Brutal Mexican drug gang crosses into U.S.," *Washington Times*, April
19, 2011, http://www.washingtontimes.com/news/2011/apr/19/violent-mexican-drug
-gang-expands-into-us/.

64. Bob Unruh, "Government calls buying 'night flashlights,' making 'extreme religious
statements' indicators of terrorism," *WorldNetDaily*, August 12, 2011, http://www.wnd
.com/index.php?fa=PAGE.view&pageId=332925.

65. "Right-wing Extremism: Current Economic and Political Climate Fueling Resurgence
in Radicalization and Recruitment," Department of Homeland Security, http://www
.fas.org/irp/eprint/rightwing.pdf.

Chapter 11. Crushing Obama's Cadre Before They Crush Us

1. Milton Mayer, *They Thought They Were Free*, Democratic Underground, January 11,
2005, http://www.democraticunderground.com/discuss/duboard.php?az=view_all&
address=103x97093.

2. Nicholas Ballasy, "Obama says he'll be taking 'executive actions' without Congress
on 'regular basis' to 'heal the economy,'" *The Daily Caller*, October 24, 2011, http://
dailycaller.com/2011/10/24/obama-says-hell-be-taking-executive-actions-without
-congress-on-regular-basis-to-heal-the-economy/.

INDEX